Polyhedra: A New Approach

by

Bruce R. Gilson

2014

Copyright 2012, 2014 by Bruce R. Gilson
All rights reserved.
Updated January 19, 2014.

Suggested library classification:
Dewey: 516.156
Library of Congress: QA491

Other books by the same author:

Construction of Musical Scales: a Mathematical Approach (2008)
Units and Measurement Systems (2008)
The Fibonacci Sequence and Beyond (2009)

All available from the same publisher.

TABLE OF CONTENTS

Preface..i

Chapter 1. Basic concepts of symmetry..1

Chapter 2. Polygons: the building blocks...12

Chapter 3. General facts about polyhedra..18

Chapter 4. The Platonic polyhedra..19

Chapter 5. Some preliminary remarks on prisms, prismoids, and antiprisms..26

Chapter 6. Prisms, Antiprisms, and Prismoids, in more detail. [Optional]...30

Chapter 7. The Archimedean polyhedra...34

Chapter 8. Modification procedures involving polyhedra I. General considerations...............................43

Chapter 9. Modification procedures involving polyhedra II. Truncation and rectification.....................44

Chapter 10. Modification procedures involving polyhedra III. Augmentation...49

Chapter 11. Modification procedures involving polyhedra IV. Uniaxial stretching.................................51

Chapter 12. Modification procedures involving polyhedra V. Dualization (general concepts)..............54

Chapter 13. Modification procedures involving polyhedra VI. Dualization (mathematical methods).
  [Optional]...58

Chapter 14. Modification procedures involving polyhedra VII. Fusion..63

Chapter 15. Modification procedures involving polyhedra VIII. Elongation and gyroelongation.........64

Chapter 16. Modification procedures involving polyhedra IX. Parallel cropping....................................65

Chapter 17. Modification procedures involving polyhedra X. Convergence..68

Chapter 18. Classification of axially symmetric polyhedra..69

Chapter 19. Apical polyhedra with triangles at the apex I. General considerations................................71

Chapter 20. Apical polyhedra with triangles at the apex II. Mathematical details. [Optional]..............74

Chapter 21. Apical polyhedra with quadrilaterals at the apex I. General considerations.......................78

Chapter 22. Apical polyhedra with quadrilaterals at the apex II. Mathematical details. [Optional]......83

Chapter 23. Apical polyhedra with pentagons at the apex I. General considerations.............................87

Chapter 24. Apical polyhedra with pentagons at the apex II. Mathematical details. [Optional]...........94

Chapter 25. A final discussion of apical polyhedra...104

Chapter 26. Preliminary remarks on tectal polyhedra..108

Chapter 27. Uni/unigeneral tectal polyhedra I. General considerations..111

Chapter 28. Uni/unigeneral tectal polyhedra II. Mathematical details. [Optional]................................116

Chapter 29. Uni/bigeneral tectal polyhedra I. General considerations...119

Chapter 30. Uni/bigeneral tectal polyhedra II. Mathematical details. [Optional]..................................126

Chapter 31. Bi/bigeneral tectal polyhedra I. General considerations..............................128

Chapter 32. Bi/bigeneral tectal polyhedra II. Mathematical details. [Optional]...............129

Chapter 33. General notes on tectal polyhedra..133

Appendix A. Useful trigonometric formulas...134

Appendix B. Concepts of vector algebra and geometry..136

Appendix C. Coordinates of the Platonic solids I. The tetrahedron..............................142

Appendix D. Coordinates of the Platonic solids II. The octahedron..............................145

Appendix E. Coordinates of the Platonic solids III. The icosahedron............................147

Appendix F. Coordinates of the Platonic solids IV. The cube..152

Appendix G. Coordinates of the Platonic solids V. The dodecahedron..........................155

Index..157

List of Tables

Table 1: Angles of polygons (in degrees)..15
Table 2: The five Platonic solids..20
Table 3: Additional parameters for the Platonic solids. (All lengths, areas, and volumes assume edge length = 1)..24
Table 4: Values of some parameters for regular prisms with square lateral faces.........29
Table 5: Coordinates of vertices of a regular antiprism...32
Table 6: Antiprism dihedral angles..33
Table 7: Primary symmetry properties of prisms, antiprisms, and prismoids................33
Table 8: The thirteen Archimedean solids...36
Table 9: Primary symmetry properties of the Archimedean solids................................41
Table 10: Regular-polygon-faced apical polyhedra with triangles at the apex..............73
Table 11: Some quantities associated with regular-polygon-based pyramids...............76
Table 12: Apical polyhedra with pentagons at the apex..93
Table 13: The Platonic solids, considered as apical polyhedra....................................104
Table 14: Primary symmetry properties of some apical polyhedra..............................107
Table 15: The various polyhedra obtained by truncating a prism................................113
Table 16: Uni/bigeneral tectal polyhedra which also qualify as Johnson solids..........123
Table 17: Uni/unigeneral tectal polyhedra, formed from the fusion of uni/bigeneral tectal polyhedra, which also qualify as Johnson solids..125
Table 18: Primary symmetry properties of some tectal polyhedra..............................133

List of Figures

Figure 1: An unmarked square..1
Figure 2: The square, with a mark added..2
Figure 3: The result of reflecting the marked square in a vertical mirror.......................2
Figure 4: An example of sixfold rotational symmetry with "vertical" mirror planes.......9
Figure 5: An example of sixfold rotational symmetry with "horizontal" mirror planes as well as horizontal twofold axes...9
Figure 6: An example of a fourteenfold alternating axis...10
Figure 7: A "polygon" which crosses itself: a pentagram..13

Figure 8: A nonconvex polygon..14
Figure 9: An octagon with fourfold dihedral symmetry..16
Figure 10: The dual of the previously illustrated octagon..16
Figure 11: An octagon with fourfold, but not dihedral, symmetry...17
Figure 12: A view of a regular tetrahedron..21
Figure 13: A view of a regular octahedron..21
Figure 14: A view of a regular icosahedron...22
Figure 15: A view of a cube (regular hexahedron)..22
Figure 16: A view of a regular dodecahedron...23
Figure 17: A prism, in this case square..26
Figure 18: A regular heptagonal antiprism...27
Figure 19: An irregular heptagonal antiprism..28
Figure 20: The Archimedean truncated cube..36
Figure 21: The Archimedean truncated octahedron..37
Figure 22: The cuboctahedron, an Archimedean solid that is both a truncation of a cube and the
    truncation of an octahedron..37
Figure 23: A cube, with no modification procedure applied. 44
Figure 24: The cube, with a small degree of truncation...45
Figure 25: The Archimedean truncated cube..45
Figure 26: The cube, with a greater degree of truncation..46
Figure 27: The cuboctahedron produced by rectifying the cube previously illustrated....................47
Figure 28: A truncated cuboctahedron, close to Archimedean, but not exactly so............................48
Figure 29: A cube, before any stretching...51
Figure 30: The prism obtained from the cube of the previous figure by a uniaxial stretching.........51
Figure 31: A dodecahedron, before any stretching..52
Figure 32: The dodecahedron from the previous figure, stretched along a fivefold axis..................52
Figure 33: A pentagonal 4-pyramoid...61
Figure 34: A regular hexagonal antiprism..66
Figure 35: A monocropped hexagonal antiprism, derived from the previously illustrated object...66
Figure 36: A symmetrically bicropped hexagonal antiprism, produced from the previous object by a
    second parallel cropping..67
Figure 37: A view of a pentagonal 4-pyramoid...79
Figure 38: An example of a polytimoid, shown with the apex downward..80
Figure 39: The dual of the polytimoid illustrated earlier in this chapter...81
Figure 40: A right hexagonal 5-pyramoid..87
Figure 41: An acute hexagonal 5-pyramoid, resembling the previous one..88
Figure 42: An obtuse hexagonal 5-pyramoid, resembling the previous one.......................................88
Figure 43: A symmetrical hexagonal orthobipyramid...89
Figure 44: The peritruncated bipyramid obtained by fusing two right hexagonal 5-pyramoids and merging
    triangles into rhombi..90
Figure 45: A hexagonal gyrobipyramid...90
Figure 46: A bicropped heptagonal antiprism..91
Figure 47: A heptagonal globoid..91
Figure 48: A heptagonal pentagonized globoid..92
Figure 49: A square 6-pyramoid...104
Figure 50: A cuboctahedron stretched along one of its twofold axes...105
Figure 51: A truncated pyramid (or apically truncated 5-pyramoid, or basally truncated prismoid)......110
Figure 52: A heptagonal antiprism..111
Figure 53: A square prism..111

Figure 54: A hexagonal prismoid 111
Figure 55: A square converged antiprism 111
Figure 56: A rectified pentagonal prism 112
Figure 57: A Johnson square cupola 120
Figure 58: A square cupola, more generally defined 120
Figure 59: An example of a monocropped hexagonal antiprism 121
Figure 60: A square rotunda 122
Figure 61: A pentagonal quasiprism 128

Preface.

I have been thinking about writing this book for a long time; this is a subject that has interested me for many years, and I have been somewhat dissatisfied with the treatments of the subject that I have seen in other books. However, the last barrier to my beginning to write this book was removed when I saw some software I needed to generate the pictures that I knew I would need to illustrate it. And so, in the last few days of July, 2011, I was finally able to begin writing the book. However, because of other demands on my time, it has taken over a year to complete it.

This book is aimed at two audiences. The first is composed of general readers who have retained enough of their high school geometry to know what, for example, an isosceles triangle is. The only additional mathematics needed will be a smattering of symmetry concepts which is covered in Chapter 1. The other is composed of those readers who would like to dig deeper into the mathematics. For them I have included "optional" chapters and appendices covering the additional material. They should look at the two Appendices describing trigonometric formulas and vector concepts used in the optional chapters, and if they do not already know the material in these Appendices, read them thoroughly, before reading the optional chapters. These chapters marked "optional" and the appendices can all be omitted if the reader desires, however.

The title of this book includes the words "a new approach." What I think is new is the emphasis on symmetry and especially a willingness to treat some polyhedra which, because they do not have regular polygons as faces, are often ignored in other works on polyhedra, but which possess symmetries that make them, in my belief, worth considering. It is also the case that, while in particular the prisms and antiprisms (and their duals) are often considered as infinite families, in this book the approach has been taken of considering as many as possible of the polyhedra in such families.

An important part of this "new approach" is classifying the various polyhedra into families based on symmetry considerations. Because this is so important, an early exposure of the reader to concepts of symmetry is essential, and the very first chapter of the book, occurring before even the concept of a polyhedron is introduced, is devoted to these concepts. The symmetry properties of both two-dimensional and three-dimensional objects are considered, because it is sometimes useful to consider the symmetries of the individual (two-dimensional) polygons that form the faces of a polyhedron as well as the symmetry of a polyhedron itself. It is, as just stated, essential that the reader familiarize himself with the concepts in this chapter before proceeding to the rest of the book, though if he is already familiar with the concepts of symmetry transformations and the nomenclature of the transformations and the symmetry groups involved (only two-dimensional and three-dimensional point groups, not any others), this chapter can probably be skipped.

One of the major contributions that was made by the mathematician Norman Johnson was his introduction of new nomenclature for polyhedra. However, while traditionally the terms "prism" and "antiprism" were defined in a way that applied to infinite families of polyhedra, Johnson defined such new terms as "cupola" and "rotunda" in a much more limited way. In this book, many of Johnson's new terms are used in a more inclusive manner, which permits them to be, as well, applied in a natural way to infinite families of polyhedra. Like Johnson, I find it necessary to introduce some new terminology, such as "polytimoid" and "globoid." One term I am *not* happy to introduce is "pentagonized globoid," however. I think a much simpler term is called for — this is, after all, a generalization of the Platonic dodecahedron by simply allowing a rotation axis to be other than threefold — and while I had considered calling it a "tetrapent" (because it had four pentagons times the order of the principal rotation axis), I thought that term ugly, and have not been able to come up with a better name. I hope that someone else will give it a better name than "pentagonized globoid."

What should be noted here is that *axial symmetry*, rather than *facial regularity*, is the characteristic that will determine whether a polyhedron is worth considering in this book. So many of the Johnson solids will be omitted from consideration. On the other hand, there are many polyhedra which cannot be constructed with regular polygon faces, and yet are of interest because there is a whole family of related polyhedra, differing only in the order of the principal rotation axis, which can be described all at one fell swoop. These *will* be discussed in this book. In fact, it was originally intended that this book be titled *Polyhedra with Axial Symmetry*, but it was decided that putting this in the title might scare away potential readers who did not know what *axial symmetry* meant.

Because of this difference of approach, I believe that this book fills a need. There are some excellent books already in circulation that deal with polyhedra, and just to recommend a pair, there are Peter R. Cromwell's *Polyhedra* and Anthony Pugh's *Polyhedra: a Visual Approach*. Each of these is worth reading, and covers many topics that this book does not, but in the same way this book covers a number of topics that neither Cromwell nor Pugh has seen fit to treat. Cromwell's book, unfortunately, is very poorly organized, however.

As I said, the final impetus toward writing this book came when I saw some software I needed to generate the pictures that I knew I would need to illustrate it. So I must give some thanks to the writers of that software: Adrian Rossiter, whose contribution is a collection of programs collectively named *Antiprism*, available for downloading on the Internet, and the person or persons, whose name(s) I do not know but who operates under the corporate name "Persistence of Vision," who wrote the ray-tracing software called *POV-Ray*. I must also thank Adrian Rossiter for explaining the usage of his software to me when I had difficulty in getting the results I wanted.

Preface.

I hope that the readers of this book will find it enjoyable, and if they wish to contact me about its content will e-mail me at brg1942@gmail.com.

Bruce R. Gilson
October 4, 2012

Note: This version incorporates a revision on January 19, 2014.

Polyhedra: A New Approach

Chapter 1.   Basic concepts of symmetry.

Before beginning anything else in this book, it is desirable to introduce some concepts of symmetry and some terminology. The definition of *symmetry* in this book will look rather different from that in other books, Web sites, or other places you might read, but it will be in fact equivalent to most definitions. This author simply finds it easier to think of symmetry in these terms, and so this will be the definition used in this book:

Suppose an object (which may be two-, three-, or in general n-dimensional for any n) to have a mark put on it, temporarily. Then a transformation will be considered as a change that could be seen as a change in the marked object, and a symmetry transformation of the object is a transformation which cannot be seen as a change if the mark is removed. In addition, the term transformation will include doing nothing at all, termed the identity transformation, and by definition the identity transformation will be a symmetry transformation for any object.

It is understood that two transformations which cannot be distinguished from each other, no matter where a mark is placed on the object, are considered the *same* transformation, and thus any transformation which leaves the object unchanged in appearance from its original form is just an example of the identity transformation.

Thus, if one takes a square (such as Figure 1 below), and puts a mark on it, one can produce an object such as is illustrated in Figure 2. (The mark has been deliberately placed in an off-diagonal position to facilitate this discussion; if it were on the diagonal, or on a line through the midpoints of opposite sides, some of the transformations to be discussed would not produce visible changes in the square. It is best to pick a place for the mark in a random location such as in Figure 2.)

Figure 1: An unmarked square.

Looking at Figure 2, it is obvious that if this square is rotated about its center by 90° (clockwise or counterclockwise; these are *two different* transformations) or 180° (in which case clockwise or counterclockwise would not matter; these rotations are equivalent), the marked square looks *different*, but performing the same operation on the original square in Figure 1 leaves it *unchanged* in appearance; thus these rotations are symmetry transformations of the square.

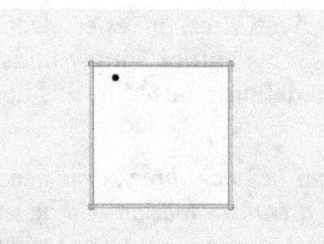

Figure 2: The square, with a mark added.

We now want to consider a different transformation. Consider a line drawn vertically through the center of the square. Now set up a transformation such that every point in the square to the *left* of that line is replaced by the point exactly the same distance to the *right* of this vertical line, and every point to the *right* of that line is replaced by the point exactly the same distance to the *left* of this vertical line. This converts the marked square of Figure 2 into the one in Figure 3, obviously different. But the original square of Figure 1 would be unchanged. So this operation is *also* a symmetry transformation of the square.

Figure 3: The result of reflecting the marked square in a vertical mirror.

When a line, such as the vertical line used to transform the marked square of Figure 2 into the one in Figure 3, is employed in this way (interchanging points on one side by those on the other side at the same distance), it is termed a *mirror line*, and the operation of interchanging the points as was done is termed *reflection (in a line)*. A mirror line is a *mathematical* concept, and thus idealized; unlike a real mirror in the world,

Chapter 1. Basic concepts of symmetry.

it reflects objects on *both* sides of it, as if, when you had a mirror in the room, objects behind it were imaged in front of it as well. Now it can easily be seen that there are three other lines that have this property, that when the marked square of Figure 2 is reflected in these lines, the appearance changes, but when the original square of Figure 1 is reflected, it is unchanged in appearance, and these are:

1. A horizontal line through the center of the square,

2. A diagonal line through the center and the upper left and lower right corners of the square, and

3. A diagonal line through the center and the lower left and upper right corners of the square.

These three lines, and the vertical line previously discussed, are the only mirror lines, reflections in which are symmetry transformations of the square; the three rotations (by 90° clockwise, by 90° counterclockwise, and by 180° in either direction) and the identity transformation, combined with the reflections in the four mirror lines, are the only eight symmetry transformations of a square.

Two-dimensional objects which are repeated in the plane (like a checkerboard pattern extended to the entire plane) can have other symmetry transformations (which will not be described in this book, but will be treated in another of my books, *Symmetry in Two Dimensions*, to be published by the same publisher as this one), but those which are not repeated can have only these three types of symmetry transformations: reflections in a line, rotations about a point, and the identity transformation, which is in a category of its own. The symmetry transformations of a given object, whether two-dimensional or three-dimensional, taken all together, form a mathematical structure called a *group*, but the mathematical theory of groups will not be undertaken in this book, although it underlies the serious study of symmetry. The only reason for introducing the concept of a group in this chapter is that it will occasionally be necessary to speak of the symmetry group of a particular polyhedron, and so the reader must understand that a *symmetry group* is really a short designation of the *totality of the symmetry transformations* of a given (two-dimensional or three-dimensional) object, that there exist names for the different symmetry groups, which will occasionally be used in this book, and that the symmetry of an object is fully described by naming the particular symmetry group to which it belongs.

It should be understood that two-dimensional objects are considered to have *no thickness at all*, so there is no such thing as the "top" or "bottom," and a mark placed anywhere can be considered to go "through": one could consider the paper on which they are drawn as some sort of idealized version of the old-fashioned blotting paper, such that anything written on the surface goes all the way through but does not spread along the surface. Or one could imagine the drawings being on some sort of super-thin and super-transparent material, such that you cannot tell which side a mark was made on; it looks the same when you turn it over as if you had left it with the same face up, except for the reversal that comes from turning it over, producing Figure 3 from Figure 2. And thus, transformations that seem to be different, like turning the paper over and reflection in a mirror line, are really the same.

As was stated a little bit previously, all the symmetry transformations of an object form a group, and one of the properties of a group is *closure*, which means, expressed in terms of symmetry transformations, that if one performs *any* symmetry transformation on an object, and follows this by *another* one (which of course can be the same one), the resulting combination is *also* a symmetry transformation of the object. This is actually pretty obvious, because if we consider the totality of *all* symmetry transformations, and *any* transformation that leaves the object apparently unchanged is a symmetry transformation, one can see that *combining* any operations that leave the object unchanged must still leave it unchanged, and so the combination must be among the symmetry transformations of the object.

Each of the symmetry groups is given a *symbol*, so that it can be referred to. There are a number of systems that have been proposed for assigning symbols to the symmetry groups; these will be taken up in more detail in my book, *Symmetry in Two Dimensions*, which is in preparation, but three are in common use. The *orbifold notation*, devised by the American mathematician John Horton Conway, has received some support from pure mathematicians, but is almost unknown to the physical scientists who also make much use of symmetry. Crystallographers prefer a notation due to Carl Hermann, a German crystallographer, and Charles-Victor Mauguin, a French mineralogist. In fact, they frequently refer to the Hermann-Mauguin notation as the International Notation, because it is used in the International Crystallographic Tables. However, the oldest of the common notations for symmetry groups is due to Arthur Moritz Schönflies, a German mathematician. The Schönflies notation was first introduced into chemistry in order to discuss the symmetry of molecules (especially in its effect on their spectra), and is still used by molecular chemists (including theoretical chemists describing molecular structure). It therefore was the notation I became most familiar with, because theoretical chemistry is the subject in which I specialized during my education. It

Chapter 1. Basic concepts of symmetry.

also appears to be the notation most frequently seen in other books on polyhedra, and for these reasons (familiarity to *me*, and common usage in *other books on this subject*) will be the one employed in this book.

A two-dimensional object can be classified into a symmetry group, as stated earlier, on the basis of its symmetry transformations. While there are an infinite number of symmetry groups that describe two-dimensional objects, they fall into two series, designated the *cyclic* and *dihedral* groups, and symbolized *C* and *D* in Schönflies' notation. Before explaining the designations, one more definition is necessary. As was noted earlier, the square has, as symmetry transformations, rotations around a center of 90° and any multiple of 90°. (It does not matter whether we consider clockwise or counterclockwise; a clockwise rotation by 270° is the same as a counterclockwise rotation by 90°.) A rotation by 90° is termed a *fourfold* rotation, because *four* such rotations will carry the square back to its original position. It will also be designated a C_4 rotation, the subscripted 4 in this symbol designating it as fourfold. With that definition having been made, we now define the *cyclic group of order* n:

A symmetry group is termed the *cyclic group of order* n if the only symmetry transformations in the group are an *n*-fold rotation and combinations of successive *n*-fold rotations around some single center. (In the case of three-dimensional objects, we will require as well that all these rotations be performed in such a way as to leave the same plane through the center unchanged. For two-dimensional objects, of course, this requirement is unnecessary, because everything is in a single plane.) We can say that the *n*-fold rotation *generates* the group, as by performing that transformation successively, one can produce every transformation in the group. (Note, specifically, that performing the rotation *n* times produces the identity transformation, so no necessity exists for any special exception to be made to the statement that "the *only* symmetry transformations in the group are an *n*-fold rotation and combinations of successive *n*-fold rotations around some single center.")

It can be seen that any *n*-fold rotation can be defined as well as a rotation through an angle of 360°/*n* or 2π/*n* radians. Since the cyclic group of order *n* is generated by a C_n rotation, it is also symbolized by C_n. Since *n* can take on any integer as a value, there are an infinite number of cyclic groups (each of which is symbolized C_n with some specific number for *n*). It can be seen that a group consisting of the identity transformation alone can be subsumed under the definition of the cyclic group of order *n*, where *n*=1, and so this trivial group, the symmetry group of an object with no symmetry, is symbolized C_1.

The other types of symmetry groups found in two-dimensional objects are the dihedral groups, and like the cyclic groups, there is one of each order:

A symmetry group is termed the *dihedral group of order* n if the only symmetry transformations in the group are an *n*-fold rotation and combinations of successive *n*-fold rotations around some single center, and the rotation of the plane containing that object by 180° about a line in that plane, turning it over. (In the three-dimensional case, "the plane containing the object" would have to be redefined, of course; this will be dealt with later in this chapter). As the *cyclic* group of order *n* is symbolized C_n, the *dihedral* group of order *n* is symbolized D_n.

Two-dimensional objects have a very limited set of symmetry types. The cyclic and dihedral groups are the only symmetry groups that are encountered in classifying the symmetries of two-dimensional objects.

One subtle bit of terminology should be observed. If one returns to the square considered earlier, it should be clear that all the transformations that belong to symmetry groups such as C_4, D_2, and the like are symmetry transformations of the square. So one can say that "the square has C_4-symmetry," or any of the others that apply. But "the symmetry group of an object" was earlier defined as a designation of the *totality of the symmetry transformations* of the object, and thus the symmetry group of the square is unequivocally D_4. It will occasionally be useful to consider all objects (or at least all polyhedra!) that have a certain symmetry, even if they are actually more symmetrical than that; thus, "all polygons that have C_2 symmetry" includes the square, though "all polygons of C_2 symmetry" or "all polygons whose symmetry is C_2" does not. It will also be useful to say that *any* polyhedron that is produced by a particular process *has* some particular symmetry, even when *some special cases* actually have *more*. So it will be useful to distinguish *having* some particular symmetry from being *of that symmetry group*.

In what has gone before, the concept of symmetry transformations has been introduced. Closely related to this is the concept of a *symmetry element*. In the reflections and rotations described, whether in two dimensions or in three, there is some point, line, or plane that can be called the symmetry element corresponding to the symmetry transformation. It remains unchanged in the transformation, and is characteristic of it. In a reflection, it would be the mirror line (in two dimensions) or mirror plane (in three); in a rotation, the center (in two dimensions) or axis (in three) of rotation. (In two dimensions, a center of rotation is also termed a *rotocenter*, a term which will sometimes be used in this book.)

Chapter 1. Basic concepts of symmetry.

It is useful to introduce a new term at this point. If an element of a figure (that is to say, any recognizable feature in that figure) is converted to another by some symmetry transformation of the figure in question, they are said to belong to the same *transitivity class*. This concept of transitivity classes will recur in this book.

Having covered the symmetries of two-dimensional figures, it is now necessary, since polyhedra are three-dimensional, to take up the symmetry of three-dimensional figures. Three-dimensional figures differ from two-dimensional in that rotations occur around a *line*, not a *point*, and mirror reflections occur in a *plane*, not a *line*. In addition, transformations that are indistinguishable in two dimensions are quite distinct in three. For example, turning an object over and mirror reflection in its own plane cannot be distinguished in a two-dimensional object. But for three-dimensional objects, a mirror reflection and a 180° rotation are never equivalent.

Of course, for three-dimensional objects just as for two-dimensional ones, the identity transformation is a symmetry transformation for *every* object, and an object for which the identity transformation is the *only* symmetry transformation is said to be in group C_1. Since, as was just stated, an object which can be rotated 180° is different from one whose only symmetry transformation (other than the identity transformation) is a mirror reflection, two different symbols are needed. Remembering that "an *n*-fold rotation" is just another way of describing a rotation through an angle of 360°/*n*, and is symbolized by C_n, an object which can be rotated 180°, and whose only symmetry transformation (other than the identity transformation) is a 180° rotation is said to be in group C_2. But an object whose only symmetry transformation (other than the identity transformation) is a *mirror reflection* needs a different symbol for its symmetry group, and (because Schönflies was German, and the German word for "mirror" is *Spiegel*) the symbol is C_s.

A three-dimensional object can also be conceptually inverted through a point (though this cannot be done with a real object, of course, but neither can a mirror reflection be done with a real object). Each point is replaced by the one *diametrically opposite* it, the same distance away from that point. In a two-dimensional object, again, a rotation through 180° (but this time within the plane, rather than about a line through the plane) accomplishes this. But a three-dimensional object whose only symmetry transformation (other than the identity transformation) is an inversion through a point is said to be in group C_i. The three groups C_1, C_i, and C_s have such low symmetry, however, that they will not be discussed further in this book, and only higher symmetries will be considered.

As was the case for two-dimensional objects, the symbol C_n describes objects whose only symmetry transformations are rotations through an angle of 360°/n and all multiples of this angle. (One multiple would be a rotation through an angle of 360°, equivalent to the identity transformation, so one does not need to add "other than the identity transformation.") But while, for two-dimensional objects, the only other possible symmetry transformations are rotations about lines perpendicular to that line (actually, about lines passing through the plane of the object), three dimensions permit a number of other types of symmetry transformations. The symbol D_n can be best thought of, in the three-dimensional case, as describing objects whose only symmetry transformations are rotations around one axis through an angle of 360°/n and all multiples of this angle, and twofold rotations about axes perpendicular to that axis. This is the most natural equivalence to the definition given earlier for the two-dimensional dihedral groups. But unlike the two-dimensional case, a number of other symbols, besides C_n and D_n, are necessary to describe all symmetry groups to which three-dimensional objects belong.

Consider an object like the one illustrated in Figure 4 below (which, in Chapter 23, will be described as a *hexagonal gyrobipyramoid*). While it is not clear, because the object is shown in a view that obscures some parts that might show this better, this object has a sixfold rotation axis through the top and bottom points. Yet it has more than just C_6 symmetry: it can be clearly seen that the object possesses a plane (actually six planes) of mirror reflection. Conventionally, it is considered that the highest rotational symmetry axis (termed the principal axis of rotation; in this case, the *only* rotational symmetry axis) is vertical, and thus any plane through this axis is considered a vertical mirror plane. To show the presence of vertical mirror planes (always, as many as the order of the axis), a subscripted "v" is added to the "6" that indicates the order of the principal axis, and the symmetry of the object illustrated in Figure 4 below is symbolized as C_{6v}. (It might be noted that, in a two-dimensional object, the presence of a mirror *line* — of course not a *plane* — through the center of rotational symmetry implies a twofold rotation about that line, so a similar object in two dimensions would be characterized as of D_6 symmetry. But in three dimensions, C_{6v} and D_6 are two quite different symmetry classes.) While the presence of a vertical mirror plane among the symmetry transformations of a group is symbolized by simply adding the "v" to the subscript, if one actually wants to refer to this reflection, one calls it by the symbol σ_v. So a list of the symmetry transformations of the C_{3v} symmetry group would include I (the identity transformation), C_3, C_3^2 (a 240° rotation), and σ_v. (This would not exhaust the list, however. Closure requires a number of others).

Chapter 1. Basic concepts of symmetry.

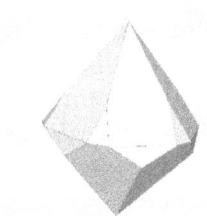

Figure 4: An example of sixfold rotational symmetry with "vertical" mirror planes.

Similarly, in the case of objects that *do* have additional twofold axes perpendicular to the principal rotation axis, and which therefore fall into the D_n family, it may also be the case that mirror planes exist. If they are horizontal, an "*h*" subscript is added to the number designating the order of the principal rotation axis, as in the object depicted in Figure 5 below (a *peritruncated hexagonal bipyramid*, in the terminology of Chapter 23).

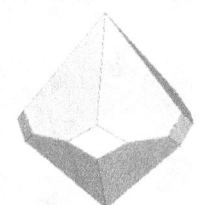

Figure 5: An example of sixfold rotational symmetry with "horizontal" mirror planes as well as horizontal twofold axes.

This object has D_{6h} symmetry, for example. There are two possible places that vertical mirror planes can occur in a D_6-type object, so one needs to distinguish between them; the "*v*" is used when the mirror planes pass through the twofold axes, and the symbol "*d*" is used when the mirror planes pass halfway between the twofold axes. (Occasionally this difference is not maintained, and D_{nv} is used to include both.) Again, the mirror reflections in the "*d*" and "*h*" planes are symbolized σ_d and σ_h when it is necessary to refer to them.

But in the three-dimensional case, an entirely different class of rotational symmetries can also exist. It is possible for an object to be such that a particular C_n rotation is not a symmetry element, and a mirror reflection in a plane perpendicular to this axis is not one either, but the *combination* of a rotation *and* a mirror reflection in a plane perpendicular to the axis of rotation *is* a symmetry element, as in the case of the object depicted in Figure 6 below. Of course, *two* such rotation-reflection combinations, performed in succession, end up as a rotation through twice the angle in question, so the illustrated polyhedron *does* possess C_7 symmetry; but the fact that a fourteenfold rotation *combined* with a reflection in a horizontal mirror is a symmetry transformation, while *neither* a C_{14} rotation nor a σ_h reflection is one, needs to be considered characteristic of this object. So a new symbol has been created to represent this combination of a C_n rotation and a σ_h mirror reflection: S_n. (Again, the S stands for the German word *Spiegel*, meaning "mirror." Given that σ is the Greek letter corresponding to "s," the same origin can be assumed for that symbol.) This symmetry transformation has been called an *improper rotation*, a term which will not be used in this book, and an *alternating axis*, which I prefer because it more accurately describes this operation. (Any point that undergoes rotation around an alternating axis, after all, alternates above and below the plane perpendicular to the axis.) And the symbol S_n (with any additional designations as necessary) will also be used to refer to a symmetry group in which the symmetry transformations include S_n but not C_n.

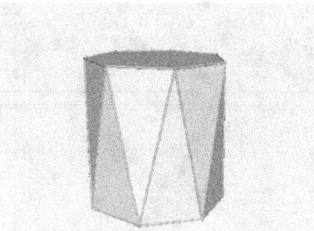

Figure 6: *An example of a fourteenfold alternating axis.*

Because the object in Figure 6 *also* possesses vertical mirror planes, its symmetry would be considered S_{14v}, although most other references would classify it differently: as D_{7d}. In those writings, only S_n symmetries, and not S_{nv}, are admitted in their classification, as it can be shown that any object with the symmetries that would classify it as S_{2nv} also has the symmetries that would classify it as D_{nd}. Because the alternating axis, leading to the *S*, is more obvious than the twofold axes that provide the *D* notation, I think that S_{14v} is a better symbol than D_{7d}. But some writers would recognize both; I have seen at least one book that used the symbol "D_{3d} (= S_{6v})," and no doubt other examples occur in the literature.

A few points should be noted about the S_n transformations:

Chapter 1. Basic concepts of symmetry.

1. Any object of C_{nh} or D_{nh} symmetry automatically has S_n as a symmetry transformation, simply because of the closure requirements. For this reason, it is unnecessary to specify the presence of an S_n axis when the group symbol is C_{nh} or D_{nh}.

2. Also because of the closure requirements, if the group symbol is S_{2n}, the symmetry transformations include C_n. This is true because two successive S_{2n} transformations produce a C_n.

3. While, in view of point 1 above, S_n *transformations* certainly exist with n odd, it is impossible for an S_n *symmetry group* to exist with n odd. For, if n is odd, n successive S_n transformations produce the same result as a σ_h mirror reflection, and $n + 1$ successive S_n transformations produce the same result as a C_n rotation, so the symmetry group is rather C_{nh}.

Besides the C, D, and S groups mentioned in this chapter, a few additional symmetry groups have been characterized and given symbols. These all have extremely high degrees of symmetry, so no one principal rotation axis can be named, and they will be covered in Chapter 4, where they are first encountered.

Chapter 2. Polygons: the building blocks.

Because the main subject of this book is *polyhedra*, but polyhedra are composed in part of *polygons*, it is necessary to discuss polygons as a preliminary to the rest of the book. It is unfortunate that the term *polygon* is used in different places to denote the polygon *boundary*, the polygon *interior*, or both, as they will shortly be defined; however, for the purposes of this book, which is mainly concerned with symmetry properties, it hardly matters which is meant; when it is necessary to refer to either the polygon boundary or the interior, this will be done explicitly, and otherwise, it should be clear that any of the three meanings can be used with equal validity.

A polygon boundary can be defined as a collection of points (the *vertices*: one is a *vertex*) and line segments (the *sides*) subject to the following conditions:

1. The number of sides and vertices is equal,
2. Every side has a vertex at each of its endpoints,
3. Every vertex is the endpoint of two sides, and
4. The entire collection is *connected, i. e.,* if one starts at any vertex, and from that vertex follows a side whose endpoint is that vertex to another, and continues from any vertex onto whichever of the two sides was *not* used to reach that vertex, ultimately every vertex is reached (or every side is traversed; both conditions are equivalent).

With these four conditions, the polygon boundary is necessarily *closed, i. e.,* one eventually reaches the starting point if one starts at any vertex and follows a path such as is described in condition 4.

This definition allows the boundary to cross itself, producing "polygons" such as the one in Figure 7. In many treatments, such boundary-self-crossing polygons are admitted, and polyhedra with these as faces are considered. In fact, there are many books and Websites devoted to these "stellated" polyhedra, including some that are named (the *Kepler-Poinsot polyhedra*). For the purposes of this book, however, no polygon will be treated whose boundary crosses itself.

Chapter 2. Polygons: the building blocks.

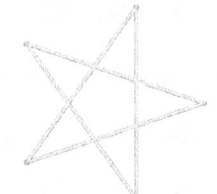

Figure 7: A "polygon" which crosses itself: a pentagram.

There is a theorem (named the *Jordan Curve Theorem*) in the branch of mathematics known as *topology* which states that a "simple closed curve" (as defined in topology) divides the plane into two regions, which can be termed the *interior* and *exterior*. As long as a polygon boundary does *not* cross itself, it satisfies the topological definition of a "simple closed curve." Thus it makes sense to speak of the interior and exterior of any polygon whose boundary does not cross itself, and since this book does not treat with boundary-self-crossing polygons, it can be stated in general that all polygons dealt with in this book possess an interior (and an exterior, about which we have nothing much to say). This will be termed the polygon interior, as mentioned at the beginning of this chapter.

Unlike the boundary-self-crossing polygons such as the one in Figure 7, the polygon in Figure 8 *does* have a distinct interior and exterior. But, because of the re-entrant (*i. e.,* greater than 180°) angle at the top center, it has the undesirable property that there are pairs of points in the upper portion that cannot be connected by a straight line, unless that line is allowed to cross the boundary (twice!) from interior to exterior and from exterior to interior. If a polygon does *not* share that property, and for *each* pair of points in the interior, they can be connected by a straight line which does *not* cross the boundary between interior and exterior, the polygon is termed *convex*. Therefore, the polygon in Figure 8 is an example of a nonconvex polygon. (A similar definition distinguishes convex from nonconvex *polyhedra*.)

Figure 8: A nonconvex polygon.

Although nonconvex polygons are not as pathological as boundary-self-crossing polygons (for example, they have well-defined interiors and exteriors), it will still be the case that the discussions in this book will mostly ignore them. In many cases, the results that are derived for polygons will apply whether they are convex or not, of course, and in some cases this will be explicitly stated.

In this book, it will be frequently desirable to refer to "isosceles polygons." Commonly, the term "isosceles" is only used of triangles and trapezoids. What these two have in common is that there is a mirror line (see Chapter 1, p. 2) which bisects a side of the polygon, and no other symmetry element. In this book, the term "isosceles" will be used of *all* polygons with a mirror line bisecting a side and no other symmetry element. The side(s) that are bisected by the mirror line are termed "base(s)" of the isosceles polygon; if the number of sides of the polygon is odd, there is one base, but if even, there are two opposite sides which are both termed bases.

It is also useful to describe a polygon with a mirror line bisecting *the angle at a vertex* and no other symmetry element. There does not seem to be a special name for this, however, a *quadrilateral* with this property, usually termed a *kite,* was also named a *strombus* by the mathematician John Horton Conway. Based on Conway's name, it seems appropriate to coin the term "*strombic polygon*" for a polygon of any number of sides with a mirror line bisecting the angle at a vertex and no other symmetry element. It should be noted that an isosceles polygon of an odd number of sides is also strombic, but an isosceles polygon of an even number of sides is not the same as a strombic polygon of that number of sides. In an odd-number-sided polygon which is isosceles (and thus also strombic) the vertex opposite the base is termed the apex of the polygon.

In elementary geometry courses, it is proved that the sum of the three angles of any triangle is 180°, and by drawing lines, it is possible to divide any polygon up into triangles, and by adding up the angles, it can be seen that if the polygon has n sides, the sum of its angles is $(n - 2)$ times 180°. If all of the angles are equal (an *equiangular* polygon) those angles will each be equal to $(1 - 2/n)$ times 180°. These values are tabulated

Chapter 2. Polygons: the building blocks.

in Table 1 below. (Since a regular polygon is a special case of an equiangular one, the third column applies to regular polygons as well.)

Number of sides	Sum of angles	Each angle, if equiangular
3	180	60.000
4	360	90.000
5	540	108.000
6	720	120.000
7	900	128.571
8	1080	135.000
9	1260	140.000
10	1440	144.000
11	1620	147.273
12	1800	150.000
13	1980	152.308
14	2160	154.286
15	2340	156.000
16	2520	157.500
17	2700	158.824

Table 1: Angles of polygons (in degrees).

Although most of the polygons with which we will be dealing will be regular, or at least strombic or isosceles (if odd-order, those are the same), it will be useful to introduce a special notation useful in classifying some of the polygons that will be considered in this book.

First, consider an octagon such as the one in Figure 9 below. Although all eight of its vertices are equivalent by symmetry, there are four sides that are all equivalent, and the other four are also equivalent, though not to the first four. The symmetry is a fourfold dihedral symmetry, notated as D_4. Although all eight vertices are equivalent, they can be put into two sets which are like left and right hands; they can only be interchanged by mirror reflection or some other operation that reverses the left and right sides in the plane. So if one only considers *rotations* in the plane, the symmetry is that of a square, reduced from that of a regular octagon. For this book, if one is concerned with the symmetry properties of a polygon, a polygon such as the one in Figure 9 is termed a 4×2-gon. The first number refers to the symmetry, which is fourfold; the second can be interpreted as indicating that if one takes a consecutive set of two vertices or two sides, that fourfold rotation will generate the entire polygon.

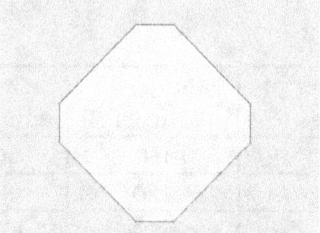

Figure 9: An octagon with fourfold dihedral symmetry.

Now consider Figure 10 below. This is a polygon related to the one in Figure 9; in fact, it was produced by locating a vertex at the midpoint of each side of the octagon in Figure 9. This is an example of a process called *"dualization"* which we will consider in more detail as it is applied to three-dimensional objects in Chapters 12 and 13. But right now, it should be noted that the symmetries of the two polygons are identical, D_4, although in Figure 9 all the *vertex angles* were equal and the *sides* were two different sizes, while in Figure 10 all the *sides* were equal and the *vertex angles* were two different sizes. In either case, a set of two vertices and two sides, rotated about a fourfold center, generates the entire octagon. So the polygon illustrated in Figure 10, too, is a 4×2-gon.

Figure 10: The dual of the previously illustrated octagon.

It should be noted that in each case, one could draw two radii from the center of the octagon, 90° apart, and by rotating about a fourfold center, generate the entire octagon. If it happened that one radius passed through a vertex, the other would as well, and the two really should be counted as a single vertex when counting the two vertices and two sides that generate the octagon. Similarly, if one radius crossed a side, so would the other, and each of those sides would only be partly in the figure in question, but the two pieces would constitute a whole side, and again, these two pieces really should be counted as a single side when counting the two vertices and two sides that generate the octagon.

Chapter 2. Polygons: the building blocks.

Finally, consider the octagon illustrated in Figure 11 below. It has two different side lengths and two different vertex angles. With both the sides and the angles split into two different sizes, the symmetry is different from the octagons in Figures 10 and 11, but still fourfold; it no longer has the mirror symmetries, and thus is now C_4 instead of D_4.

Figure 11: An octagon with fourfold, but not dihedral, symmetry.

However, for this notation, all that matters is the fact that there are four sets of two sides and 2 vertices, any one of which generates the entire octagon by fourfold rotation. So the polygon illustrated in Figure 11, too, is a 4×2-gon.

While Figures 9, 10, and 11 are all $n\times2$-gons (specifically, 4×2-gons, in fact), in nearly all the cases where the term "$n\times2$-gon" is used in this book, it will refer to one like the one illustrated in Figure 9, where the angles are all equal (and thus, equal to the vertex angles of a regular $2n$-gon), and the sides alternate in length. This should be assumed, in the absence of any other descriptive information, in all subsequent places where this term is used.

Using this notation, a regular n-gon is an $n\times1$-gon, while a polygon with no rotational symmetry (which does not mean no symmetry at all; it might be isosceles or strombic) is a $1\times n$-gon. An $m\times n$-gon is of necessity an mn-gon, whatever the values of m and n, but characterizing a polygon as an mn-gon says nothing about the symmetry, while specifying that it is an $m\times n$-gon implies m-fold symmetry, either C_m or D_m. It does *not*, however, imply that the sides or the vertices fall into n different transitivity classes. It can be seen that the vertices in Figure 9, and the sides in Figure 10, fall into a single transitivity class.

Chapter 3. General facts about polyhedra.

A polyhedron, like a polygon, may be defined in more than one way, and the term sometimes means the boundary and sometimes the interior, or even both together, which is the usual meaning in this book. Just as in the case of polygons, however, it hardly matters which is meant; when it is necessary to refer to either the polyhedron boundary or the interior, this will be done explicitly, and otherwise, it should be clear that any of the three meanings can be used with equal validity.

The boundary of a polyhedron consists of a number of polygons, called the *faces* of the polyhedron. Every side of a polygon in the polyhedron boundary is also the side of exactly one other polygon, and these shared sides are called edges of the polyhedron. No two polygons which share an edge may be in the same plane. Connectedness requirements, similar to those for polygons, are imposed. The vertices of the polygons are also termed vertices of the polyhedron.

At any vertex of a polyhedron, at least three faces must come together. Consider one face that includes that vertex. Like all polygonal vertices, two sides of the polygon under consideration must come together at that vertex. Each of those sides must be shared with another face of the polyhedron, and those must be two distinct faces because any figure that includes both of those line segments must be in the same plane as the first face. But each of these faces must, since they share sides with the original face, share the endpoints of those sides, which include the common vertex that started this discussion. That completes the proof, and similarly it can be shown that at any vertex of a polyhedron, at least three edges must come together.

The three (or more) faces that come together at a vertex each have an angle at that vertex, and if the sum of those angles were 360°, all those polygons would be in a plane. If it were greater than 360°, the polygons could not be fitted together (try it, if you do not believe this!). Thus, the vertex angles of all the faces coming together at a vertex must have a sum which is *strictly less* than 360°. The difference between this sum and 360° is called the *angle defect* of the vertex, and the total of the angle defects of all the vertices of a well-behaved polyhedron (that is, not possessing holes through it or falling into disconnected pieces) will always be 720°. (This was shown by René Descartes centuries ago.) Because the *total* of the angle defects of *all* the vertices is 720°, it is clear that the *average* of the angle defects of all the individual vertices will become smaller as the number of vertices increases. Now, if the polygons at a vertex were all in one plane, the angle defect at the vertex would be *zero*. And this, in turn, means that, in a polyhedron with a large number of vertices, each vertex is more nearly planar than in a polyhedron with a smaller number of vertices.

Chapter 4. The Platonic polyhedra.

To be termed "regular" a *polygon* needs to meet only two criteria:

1. All its sides must be equal in length, and
2. All the angles at its vertices must be equal.

By contrast, *polyhedra*, to be classified as regular, are required to meet a larger number of criteria, although they can be expressed as well as two, because the first, in particular, contains a number of subcriteria which need to be obeyed to meet it:

1. All the faces must be regular polygons of the same kind (*i. e.*, all triangles, all squares, etc.), and
2. All the vertices must have the same number of polygons meeting.

Because the faces are regular and of the same kind, the first criterion actually calls for all edges to be equal, all vertex angles to be equal, and more, so the apparent two criteria in fact set a much larger number of conditions.

Combining the known values of the vertex angles of a regular polygon (from Table 1) with the requirement that the vertex angles of all the faces coming together at a vertex must have a sum which is less than 360° (See p. 18 in Chapter 3), it can be seen that only *five* are possible. Because the Greek philosopher Plato, in his work *Timaeus*, described all five of them, these regular polyhedra are usually termed the Platonic solids or Platonic polyhedra.

First of all, since at least three faces must meet at each vertex, and all the polygons are regular, the only polygons that can be faces are triangles, squares, and pentagons. From Table 1, three hexagons would have an angle sum of exactly 360°, and three polygons of any larger number of sides would exceed 360°.

Second, since four squares would have an angle sum of exactly 360°, and four pentagons would exceed that figure, there can be only exactly three of each of those two types of polygon. But with triangles, it would take six to equal 360°, so any number less than six is acceptable. This gives us the five possibilities: three, four, or five triangles, three squares, and three pentagons.

Of course, all that has been demonstrated so far is that *if* regular polyhedra *can* be constructed, those five possibilities are all that can be expected. It does not prove that such regular polyhedra *actually can be constructed;* that was demonstrated by the ancient Greeks, however, and, for example, Euclid's *Elements* contain demonstrations of the construction of each of them. They are listed below in Table 2.

Number of vertices of each polygon	Number of polygons meeting at a vertex	Total number of faces	Total number of edges	Total number of vertices	Name
Number of sides of each polygon	Number of edges meeting at a vertex	F	E	V	
3	3	4	6	4	Tetrahedron
3	4	8	12	6	Octahedron
3	5	20	30	12	Icosahedron
4	3	6	12	8	Hexahedron or cube
5	3	12	30	20	Dodecahedron

Table 2: The five Platonic solids.

It might be observed that all these polyhedra obey the *Euler formula* $V + F = E - 2$. In fact, unless a polyhedron has a hole or consists of several pieces, it will always obey this formula. (There is a connection between the Euler formula and the fact that the sum of the angle defects at all the vertices is 720°, mentioned on p. 18 previously. In fact, if one calculates the number $V + F - E$ for any structure of plane faces, whether it has holes or consists of several pieces, it will always be the case that the total defect will be $360°[V + F - E]$.)

The requirement that the sum of the angle defects at all the vertices is 720° can be seen to have some plausibility when one notices that the smaller the angle defect, the more nearly planar the portion of the polyhedron around that vertex is. If the angle defects are small, then, there will be more vertices needed to close up the polyhedron. This does not prove that the angle defects must add to exactly 720°, but it does make it seem that small angle defects mean lots of vertices and large ones mean fewer vertices. (As stated on p. 18, the rule was proved by René Descartes, so it is a long-known rule, however.)

Strictly speaking, any polyhedron with four faces can be termed a *tetrahedron*; any polyhedron with six faces can be termed a *hexahedron*; any polyhedron with eight faces can be termed a *octahedron*; any polyhedron with twelve faces can be termed a *dodecahedron*; and any polyhedron with twenty faces can be termed an *icosahedron*. When it is specifically intended to indicate the five *Platonic solids*, the word "regular" should be added before these terms. However, as in a chapter like this one, specifically dealing with the Platonic solids, the term "regular" will be omitted and assumed to be

Chapter 4. The Platonic polyhedra.

implied.

The tetrahedron is the simplest polyhedron possible. No polyhedron is possible with fewer than four faces, or fewer than six edges, or fewer than four vertices. As stated in the previous paragraph, any polyhedron with four faces can be termed a *tetrahedron,* but in fact all tetrahedra will have four triangular faces, and four vertices, at each of which three of the four faces meet. Every pair of faces selected from the four shares an edge, making six in all, and so the only difference between an irregular tetrahedron and a regular one is that all the edges of a regular tetrahedron are equal. One can consider any irregular tetrahedron as a distorted regular tetrahedron. A regular tetrahedron is pictured in Figure 12 below.

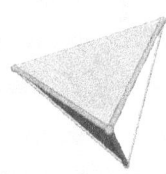

Figure 12: A view of a regular tetrahedron.

Unlike the tetrahedron, for all of the *other* Platonic solids, there are polyhedra with the same numbers of faces that cannot be thought of as distorted versions of the corresponding Platonic solid. For example an octahedron can be a pyramid with a heptagonal base, or any of a number of other types of polyhedron. But a *regular* octahedron looks like the object pictured in Figure 13 below. (However, in this book, "octahedron" will normally *mean* "regular octahedron.")

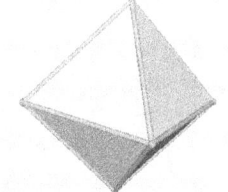

Figure 13: A view of a regular octahedron.

Everything that was said about the octahedron in the previous paragraph could be equally well applied to the icosahedron. The number of twenty-faced polyhedra that can be constructed that are essentially different, not merely distortions of one another, is probably so large that it has not been determined for a fact. But in this book, the term "icosahedron" will mean a *regular* icosahedron, such as illustrated in Figure 14, exclusively.

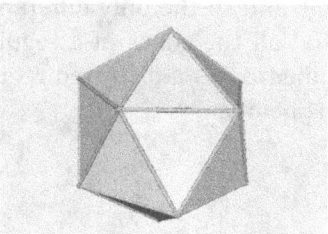

Figure 14: A view of a regular icosahedron.

Because the regular hexahedron has an alternate name (cube) it is not as common to omit the word *"regular"* and simply call a regular hexahedron a *hexahedron* without the qualifying adjective. The word *"cube"* is even shorter, and so is the word of choice. Although most readers will have a clear idea what a cube is, it is illustrated in Figure 15 below.

Figure 15: A view of a cube (regular hexahedron).

There is actually another twelve-sided figure, besides the regular dodecahedron, which often gets the name "dodecahedron." It is a figure composed of twelve rhombi, with some vertices at which three rhombi meet, and others at which four meet. Because the faces are all rhombi, it is termed a *rhombic dodecahedron*, and has some importance in crystallography, as well as being related to some of the other polyhedra which *do* have regular polygons for faces. But when the term "dodecahedron" is used, it normally refers to a *regular* dodecahedron, sometimes called a *pentagon-dodecahedron* to distinguish it from a rhombic dodecahedron. An illustration of this polyhedron is in Figure 16 below.

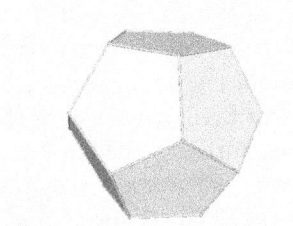

Figure 16: A view of a regular dodecahedron.

It can be seen that the octahedron and cube have a relationship such that the first column of the listing of either one is the second column of the listing of the other one, and the third column of the listing of either one is the fifth column of the listing of the other one. The icosahedron and dodecahedron have the same relationship, but attempting to make the same interchanges of columns for the tetrahedron, one is returned to the same numbers in the same columns. The term *duality* is applied to this kind of relationship (we say the cube is the *dual polyhedron* to the octahedron, and vice versa), and more will be said about this in later chapters as well. Of course, for other than the Platonic solids, the "Number of sides of each polygon" would not necessarily be the same for each polygon, so the precise definition of duality would need to be made more explicit for such polyhedra. But the duality relationships for the Platonic solids are clear:

1. The *octahedron* and *cube* are duals,
2. The *icosahedron* and *dodecahedron* are duals, and
3. The *tetrahedron* is *its own* dual (or, the tetrahedron is *self-dual*).

This will be returned to later.

In Chapter 1, the symmetries discussed, denoted by C_n, D_n, and S_n, with additional subscripted letters for mirror planes, all presupposed one single principal rotation axis. The Platonic solids have much higher symmetries, and special symbols are used for these higher symbols. The regular tetrahedron, for example, has four different axes, each of which shows all the symmetries of the C_{3v} group. In addition, it has three twofold (actually alternating fourfold) axes, but unlike the twofold axes appearing in the specification of D_n symmetry, they do not run perpendicular to any of the threefold axes. So a special symbol, T_d, is used to denote the symmetry of the tetrahedron; if some object has all the symmetries of the tetrahedron except its mirror planes, the symmetry is symbolized T, without the subscripted "d." The regular octahedron and cube share a common symmetry. Each has four different threefold (actually alternating sixfold) and three different fourfold axes, each of which has mirror planes through it. Treating any one of the fourfold axes as a principal rotation axis would give it D_{4h} symmetry, and treating any of the alternating sixfold axes as a principal rotation axis would give it S_{6v} symmetry, but there is no reason to favor any one of these axes (although, as we will see later, many modification procedures applied the octahedron or cube reduce the symmetry to S_{6v}, D_{4h}, or even lower) so a special symbol, again, is adopted: O_h. (The letter C for *cube* has already been taken, so O, for *octahedron*, is the symbol of choice; if some object has all the symmetries of the octahedron and cube except their mirror planes, the symmetry is symbolized O, without the subscripted "h.") And the dodecahedron and icosahedron also share a common symmetry, with ten different threefold (actually alternating sixfold) and six different fivefold (actually alternating tenfold) axes, each of which has mirror planes through it. Again, there is no reason to favor one of these axes, so a special symbol again is adopted, I_h. (Occasionally, the letter Y is used instead of I, for unknown reasons.) And again, the subscript "h" is omitted if the mirror planes are absent in an object that otherwise shares the symmetries of the icosahedron and dodecahedron.

In add to the information in Table 2, there are a number of other quantities that characterize the Platonic solids, and the most important are given in Table 3 below. All the quantities (except for angles) depend on the edge length (or other quantity determining the scale, such as a height), and are shown for edge length = 1. If the edge length is $e \neq 1$, all lengths are to be multiplied by e^2, and the volume by e^3.

name(s)	total surface area	surface area of a face	volume	Dihedral angle	central angle	circumradius	inradius	intersphere radius
tetrahedron	1.732051	0.433013	0.117851	70° 32'	109° 28'	0.612372	0.204124	0.353553
octahedron	3.464102	0.433013	0.471405	109° 28'	90°	0.707107	0.408248	0.500000
icosahedron	8.660254	0.433013	2.181695	138° 11'	63° 26'	0.951057	0.755761	0.809017
hexahedron (cube)	6.000000	1.000000	1.000000	90°	70° 32'	0.866025	0.500000	0.707107
dodecahedron	20.645729	1.720477	7.663119	116° 34'	41° 49'	1.401259	1.113516	1.309017

Table 3: Additional parameters for the Platonic solids. (All lengths, areas, and volumes assume edge length = 1)

The *dihedral angle* is the angle between the planes of any two faces which share an edge. The *central angle* is the angle between any two radii (line segments from the

Chapter 4. The Platonic polyhedra.

centroid of the polyhedron to a vertex of the polyhedron) which are adjacent (i. e., two ends of the same edge). All the vertices of a Platonic solid, because of the symmetry, lie on a single sphere, the *circumsphere*, and its radius is termed the *circumradius*. All the faces of a Platonic solid, again because of the symmetry, are tangent to a single sphere, the *insphere*, and its radius is termed the *inradius*. The *intersphere* (also termed the *midsphere*) is a sphere tangent to every one of the edges of the polyhedron; not every polygon has an intersphere, but highly symmetrical ones, such as the Platonic solids, do.

Of these quantities tabulated in Table 3, the one that will be most important in this book is the dihedral angle, which will turn up in a number of places, nowhere more important than in Chapter 14.

It should be noted that all the vertices of a Platonic polyhedron fall into a single transitivity class. If any polyhedron has the property that its vertices all fall into a single transitivity class, it is said to be *isogonal*; if its vertices fall into n different transitivity classes ($n>1$) it is said to be n-*isogonal*. (In some literature, the more transparent term *vertex-transitive* is used rather than *isogonal,* but no term equivalent to n-*isogonal* exists in these writings.) Similarly, all the faces of a Platonic polyhedron fall into a single transitivity class. If any polyhedron has the property that its faces all fall into a single transitivity class, it is said to be *isohedral*; if its faces fall into n different transitivity classes ($n>1$) it is said to be n-*isohedral*. (In some literature, the more transparent term *face-transitive* is used rather than *isohedral,* but no term equivalent to n-*isohedral* exists in these writings.) And finally, all the edges of a Platonic polyhedron fall into a single transitivity class. If any polyhedron has the property that its edges all fall into a single transitivity class, it is said to be *isotoxal*; if its edges fall into n different transitivity classes ($n>1$) it is said to be n-*isotoxal*. (In some literature, the more transparent term *edge-transitive* is used rather than *isotoxal,* but no term equivalent to n-*isotoxal* exists in these writings.) The Platonic polyhedra and two of the Archimedean polyhedra (defined in Chapter 7) are virtually the only isotoxal polyhedra that one will encounter, but isogonal and isohedral polyhedra will be encountered in subsequent chapters.

It might be noted that in tables occurring in this book, "the value of n for which a type of polyhedron is n-isogonal" (or n-isohedral, or n-isotoxal) may be given as 1. Of course, based on the definition above, "1-isogonal" simply means the same as isogonal, as also "1-isohedral" simply means the same as isohedral, and "1-isotoxal" simply means the same as isotoxal; the usage of the terms with the "1" will only be in such tabulations where a value of n is called for.

Chapter 5. Some preliminary remarks on prisms, prismoids, and antiprisms.

Although the logical place to discuss prisms, prismoids, and antiprisms in this book is later, after some concepts that are helpful in discussing them have been developed, a short preliminary chapter needs to be included at this point, because there will be continual reference to prisms, prismoids, and antiprisms all through the remainder of this book. So they will be defined here, but the bulk of the discussion will be postponed until later chapters.

A *prism* is a polyhedron which has two faces (usually called *bases*) that are located in parallel planes and which are congruent (identical in size and shape) polygons. The other faces (termed *lateral faces*) are rectangles located such that two opposite sides of the rectangle are corresponding sides of the two bases, while the other two sides are edges of the polyhedron connecting corresponding vertices of the bases. (Actually, what has been defined here is a *right prism*; it is also possible to have an *oblique prism,* which differs in that the lateral faces are *parallelograms,* rather than rectangles, and the edges joining corresponding vertices of the bases are *not* perpendicular to the bases, but *are* all parallel to *each other*. These will not, however, be treated in this book, so the term *prism,* when used in this book, shall be understood as meaning *right prism*.)

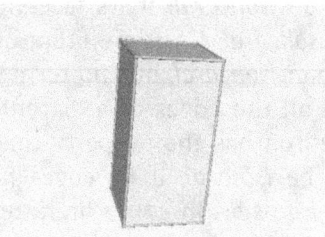

Figure 17: A prism, in this case square.

Figure 17 is an example of a prism. Since the bases are squares, the prism illustrated in Figure 17 is termed a *square prism*. It should be noted that in no part of this definition is there any requirement that the bases be regular polygons, although most of the prisms discussed in this book will have regular polygons as bases. The term *regular prism* will be used if it is specifically desired to include a requirement that the bases be regular polygons.

If the requirement that the bases be *congruent* is relaxed, but it is still required that they be *similar* (identical in *shape,* but not necessarily the same *size*), a polyhedron is obtained which has (among other names) been termed a *prismoid,* or a *frustum of a pyramid.* (However, the term "*prismoid*" has been used by some authors with different meanings, so the reader should check, if he encounters the term elsewhere, whether the same meaning is attached to the term as in this book. In this book, the term "prismoid" will be used in this sense.) Corresponding sides of the bases are still required to be parallel, as in the case of a prism, but the lateral faces are trapezoids, rather than rectangles.

An antiprism is also a polygon which has two bases which are polygons of the same type, located in parallel planes. However, they are not required to be congruent, and they must be located so that the vertices of one polygon are not directly over vertices of the other. For most of the purposes of this book, we will restrict ourselves to regular antiprisms, in which the polygons are congruent regular polygons, rotated with respect to each other by half the central angle of the polygon. (The *central angle* means the angle in the polygon between two lines from the center to the opposite ends of a single side.) With this restriction, each vertex of one base is equidistant from the nearest two vertices of the other base, and each combination of a vertex of one base and the two nearest vertices of the other form the vertices of a triangle. These triangles, the *lateral faces* of the antiprism, will number twice the number of sides of each base, because there will be one triangle joining *each* vertex of *each* base with the two nearest vertices of the other. An example of a regular antiprism is shown below in Figure 18. In this case, the bases are regular seven-sided figures (heptagons), so the antiprism is termed a heptagonal antiprism, but any regular polygon, from triangles on up, is possible. (In fact, it is possible to conceptualize a *digonal antiprism*, with each of the bases being a line segment counted twice, and which will share the properties of a regular antiprism in which the number of sides is considered to be 2.)

Figure 18: A regular heptagonal antiprism.

To the extent that antiprisms that are not regular are discussed, one requirement will be enforced: for each vertex of either base, a line passing through that vertex, perpendicular to the planes of the bases, meets the plane of the other base *outside* the polygon that constitutes that base. This makes certain that, though the bases need not be congruent, they are approximately the same size. Figure 19 below shows an example of an irregular antiprism, in this case (like Figure 18) having heptagons for bases, but irregular ones.

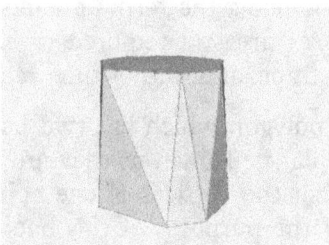

Figure 19: An irregular heptagonal antiprism.

A regular antiprism may have equilateral triangles for lateral faces, but need not; the antiprism of Figure 18, for example, has long, narrow isosceles triangles. The lateral faces will, however, at least be congruent isosceles triangles, although the proportions of those triangles are essentially unrestricted. When the antiprism is irregular, the triangles may be scalene, though occasionally some might be isosceles or even equilateral. The fact that the bases are irregular, however, will mean that the lateral faces will, in general, not be congruent.

It should be noted that if the bases of a prism or antiprism are regular polygons, the polyhedron will be isogonal. (In fact, the bases need not be regular, but merely the sort of polygon which has all of its vertices in a single transitivity class.) Similarly, a regular-polygon-based prism or antiprism is 2-isohedral (the bases falling in one transitivity class, the lateral faces into another) and 2-isotoxal, and just as was noted for the isogonality, it is possible, if the bases have the appropriate symmetry, for a prism or antiprism with polygonal bases that are not regular to be 2-isohedral and 2-isotoxal. In other words, every regular-polygon-based prism or antiprism is isogonal, 2-isohedral and 2-isotoxal, but not every isogonal, 2-isohedral and 2-isotoxal prism or antiprism has regular polygons for bases. (Note, however, that the additional symmetry of *that particular type* of square prism which is a *cube*, and *that particular type* of triangular antiprism that is a *regular octahedron*, reduces the number of transitivity classes of edges and faces to 1, and they are isohedral and isotoxal, as examples of Platonic solids!)

Unlike prisms and antiprisms, prismoids are not isogonal, because no symmetry of the polyhedron could interchange the two bases (since they are not the same size!) But when a prismoid has *regular* polygons (or *any* type of polygon which has all of its vertices in a single transitivity class) for bases, it certainly will be 2-isogonal; the vertices at *each* base will all fall into a single transitivity class. And in that case, it will be 3-isohedral, because the two bases are in different transitivity classes, but all the lateral faces are equivalent, and 3-isotoxal, for similar reasons.

In the case of the Platonic solids, a table such as Table 3 could be constructed, giving certain data that are characteristic of each. For prisms, antiprisms, and prismoids,

Chapter 5. Some preliminary remarks on prisms, prismoids, and antiprisms.

however, specific values of many of those parameters appearing in Table 3 cannot be given in general, because they depend on more than simply the number of vertices at each base. It is possible, however, to look at the specific cases where the lateral faces of the polyhedron are *regular* (*squares*, in the case of prisms; *equilateral triangles*, in the case of antiprisms). And in that case, specific values can be given. (Because, other than the case of the prisms and antiprisms that are Platonic solids, *i. e.*, the *cube* and *regular octahedron*, there is no inscribed sphere, no inradius is given.)

Table 4 below gives some parameters of the regular prisms whose bases have from 3 to 17 sides, with square lateral faces, analogous to the corresponding Table 3 for the Platonic solids. The dihedral angles are not given, because all the dihedral angles between either base and any lateral face are always 90°, and the dihedral angles between adjacent lateral faces are equal to the angle at the vertex of a regular polygon, and can thus be read from Table 1.

Number of sides of each base	Total area of faces	Volume	Circumradius	Radius of intersphere
3	3.866025	0.433013	0.763763	0.577350
4	6.000000	1.000000	0.866025	0.707107
5	8.440955	1.720477	0.986715	0.850651
6	11.196152	2.598076	1.118034	1.000000
7	14.267825	3.633912	1.256179	1.152382
8	17.656854	4.828427	1.398966	1.306563
9	21.363648	6.181824	1.545043	1.461902
10	25.388418	7.694209	1.693527	1.618034
11	29.731280	9.365640	1.843821	1.774733
12	34.392305	11.196152	1.995508	1.931852
13	39.371537	13.185768	2.148287	2.089291
14	44.669004	15.334502	2.301938	2.246980
15	50.284726	17.642363	2.456295	2.404867
16	56.218716	20.109358	2.611233	2.562915
17	62.470984	22.735492	2.766652	2.721096

Table 4: Values of some parameters for regular prisms with square lateral faces.

Chapter 6.   Prisms, Antiprisms, and Prismoids, in more detail.
[Optional]

In this chapter, we will always denote the number of sides of each base of a prism, antiprism, or prismoid by n, and it will be assumed that both bases are regular n-gons. Using a notation that will be explained more precisely in Chapter 18, the vertices of one base will be designated as $A_1, A_2, ..., A_n$, and those of the other as $B_1, B_2, ..., B_n$. The z-axis will be assumed to pass through the centers of both bases. For a prism or antiprism, it is simplest to locate the origin halfway between the two planes, because then the symmetry properties of the polyhedron are most easily seen; for a prismoid, these symmetries do not exist, but it will be convenient to do the same. So if h denotes the height of the prism, antiprism, or prismoid, i. e., the perpendicularly measured distance between the bases, the z-coordinates of all the A vertices can be written as $h/2$ and the z-coordinates of all the B vertices as $-h/2$. For a prism or antiprism, with both bases being congruent n-gons, the coordinates of the vertices can all be expressed in terms of a single radius of the circle that circumscribes either base r. For a prismoid, two different such radii are needed, and they can be denoted r_A and r_B, with the letter A or B chosen to match the letter used for the vertex. With these conventions established, the coordinates of the vertices of a prism (with j = 1, 2, ..., n) are:

A_j: $(r \cos [j - 1]\alpha, r \sin [j - 1]\alpha, h/2)$

B_j: $(r \cos [j - 1]\alpha, r \sin [j - 1]\alpha, -h/2)$, where α = $2\pi/n$ radians or $360°/n$.

The coordinates for an antiprism can similarly be established as:

A_j: $(r \cos [j - 1]\alpha, r \sin [j - 1]\alpha, h/2)$

B_j: $(r \cos [j - \frac{1}{2}]\alpha, r \sin [j - \frac{1}{2}]\alpha, -h/2)$,

and those for a prismoid as:

A_j: $(r_A \cos [j - 1]\alpha, r_A \sin [j - 1]\alpha, h/2)$

B_j: $(r_B \cos [j - 1]\alpha, r_B \sin [j - 1]\alpha, -h/2)$.

Although a prismoid, by its definition, cannot have *all* its edges equal, prisms and antiprisms can, and for those, it will be useful to determine the values of h and r necessary to produce a prism or antiprism whose edges are all equal, and which thus has only regular polygons for faces. In the case of a prism, each lateral face is a rectangle one of whose dimensions is h, so it can be made square by simply making the distance $A_1A_2 = h$, as all the other distances A_jA_{j+1} and B_jB_{j+1} (and A_nA_1 and B_nB_1) will be equal, by symmetry, as well to h. But the distance $A_1A_2 = r\sqrt{[(\cos \alpha - 1)^2 + \sin^2 \alpha]} = r\sqrt{[2(1 - \cos \alpha)]}$. So it is simply necessary to make

$$r = h/\sqrt{[2(1 - \cos \alpha)]}$$

to make all the sides equal.

Chapter 6. Prisms, Antiprisms, and Prismoids, in more detail. [Optional]

For an antiprism, the calculation is more complicated. One can no longer take h equal to the size of the edge. But the relationship can be given by first calculating the distance A_1A_2 in terms of r as before, and A_1B_1 in terms of r and h. The Pythagorean formula gives:

$$A_1A_2 = r\sqrt{[2(1 - \cos \alpha)]}, \text{ as before, and}$$

$$A_1B_1 = \sqrt{[2(1 - \cos \alpha/2)r^2 + h^2]}.$$

These must be equal, but it is simpler to work with them if instead we make $A_1A_2^2 = A_1B_1^2$:

$$[2(1 - \cos \alpha)]r^2 = 2(1 - \cos \alpha/2)r^2 + h^2, \text{ or}$$

$$(1 - \cos \alpha)r^2 = (1 - \cos \alpha/2)r^2 + h^2/2;$$

$$r^2(\cos \alpha/2 - \cos \alpha) = h^2/2;$$

$$h^2 = 2r^2(\cos \alpha/2 - \cos \alpha).$$

It is probably better, however, to express both h and r in terms of the edge that all of the polygons are to have in common. If we let e be that edge, the formula to compute the r value of the base in terms of the side has already been worked out for the prism, except that in that case the side of the base was equal to h. Allowing for this difference, we can put

$$r = e/\sqrt{[2(1 - \cos \alpha)]}, \text{ combined with the already-derived}$$

$$h^2 = 2r^2(\cos \alpha/2 - \cos \alpha), \text{ which in terms of } e \text{ can be written}$$

$$h = e\sqrt{[(\cos \alpha/2 - \cos \alpha)/(1 - \cos \alpha)]}.$$

Using the values of r and h computed by this formula and taking $e=1$, Table 5 below shows the values of the coordinates of the vertices A_1, B_1, and A_2 for regular antiprisms of 3 to 17 sides. It should be noted that for large values of n, the values of h become very close to $½\sqrt{3} = 0.8660254$, making the z-coordinates of the vertices very close to $¼\sqrt{3} = 0.4330127$. There is actually a clear reason for this: as the number of vertices increases, the lateral faces become more nearly parallel to the z-axis, so that they are equilateral triangles whose altitudes are close to being parallel to the z-axis. (This can also be noted by examining Table 6, which shows that the dihedral angle between either base and a lateral face becomes close to 90°.) An equilateral triangle with side 1 has an altitude of $½\sqrt{3}$, accounting for the limiting value of h.

Polyhedra: A New Approach

n	A_1	B_1	A_2
3	(0.57735027, 0.00000000, 0.40824829)	(0.28867513, 0.50000000, -0.40824829)	(-0.28867513, 0.50000000, 0.40824829)
4	(0.70710678, 0.00000000, 0.42044821)	(0.50000000, 0.50000000, -0.42044821)	(0.00000000, 0.70710678, 0.42044821)
5	(0.85065081, 0.00000000, 0.42532540)	(0.68819096, 0.50000000, -0.42532540)	(0.26286556, 0.80901699, 0.42532540)
6	(1.00000000, 0.00000000, 0.42779984)	(0.86602540, 0.50000000, -0.42779984)	(0.50000000, 0.86602540, 0.42779984)
7	(1.15238244, 0.00000000, 0.42923660)	(1.03826070, 0.50000000, -0.42923660)	(0.71849870, 0.90096887, 0.42923660)
8	(1.30656296, 0.00000000, 0.43014778)	(1.20710678, 0.50000000, -0.43014778)	(0.92387953, 0.92387953, 0.43014778)
9	(1.46190220, 0.00000000, 0.43076304)	(1.37373871, 0.50000000, -0.43076304)	(1.11988206, 0.93969262, 0.43076304)
10	(1.61803399, 0.00000000, 0.43119850)	(1.53884177, 0.50000000, -0.43119850)	(1.30901699, 0.95105652, 0.43119850)
11	(1.77473277, 0.00000000, 0.43151824)	(1.70284362, 0.50000000, -0.43151824)	(1.49300021, 0.95949297, 0.43151824)
12	(1.93185165, 0.00000000, 0.43176003)	(1.86602540, 0.50000000, -0.43176003)	(1.67303261, 0.96592583, 0.43176003)
13	(2.08929073, 0.00000000, 0.43194739)	(2.02857974, 0.50000000, -0.43194739)	(1.84997507, 0.97094182, 0.43194739)
14	(2.24697960, 0.00000000, 0.43209553)	(2.19064313, 0.50000000, -0.43209553)	(2.02445867, 0.97492791, 0.43209553)
15	(2.40486717, 0.00000000, 0.43221473)	(2.35231505, 0.50000000, -0.43221473)	(2.19695548, 0.97814760, 0.43221473)
16	(2.56291545, 0.00000000, 0.43231206)	(2.51366975, 0.50000000, -0.43231206)	(2.36782513, 0.98078528, 0.43231206)
17	(2.72109558, 0.00000000, 0.43239258)	(2.67476375, 0.50000000, -0.43239258)	(2.53734606, 0.98297310, 0.43239258)

Table 5: Coordinates of vertices of a regular antiprism.

Chapter 6. Prisms, Antiprisms, and Prismoids, in more detail. [Optional]

n	Dihedral angle at base		Dihedral angle between lateral faces	
	rad	deg. min. sec.	rad	deg. min. sec.
3	1.91063324	109° 28′ 16.39″	1.91063324	109° 28′ 16.39″
4	1.81228288	103° 50′ 10.18″	2.22619544	127° 33′ 5.77″
5	1.75950686	100° 48′ 44.34″	2.41186500	138° 11′ 22.87″
6	1.72612066	98° 53′ 57.94″	2.53460015	145° 13′ 18.81″
7	1.70295715	97° 34′ 20.13″	2.62187309	150° 13′ 20.15″
8	1.68589238	96° 35′ 40.27″	2.68715051	153° 57′ 44.58″
9	1.67277542	95° 50′ 34.70″	2.73783237	156° 51′ 58.46″
10	1.66236755	95° 14′ 47.92″	2.77832867	159° 11′ 11.42″
11	1.65390239	94° 45′ 41.86″	2.81143311	161° 4′ 59.71″
12	1.64687931	94° 21′ 33.24″	2.83900225	162° 39′ 46.25″
13	1.64095689	94° 1′ 11.65″	2.86231843	163° 59′ 55.56″
14	1.63589405	93° 43′ 47.37″	2.88229598	165° 8′ 36.22″
15	1.63151559	93° 28′ 44.25″	2.89960451	166° 8′ 6.36″
16	1.62769105	93° 15′ 35.38″	2.91474568	167° 0′ 9.45″
17	1.62432126	93° 4′ 0.31″	2.92810280	167° 46′ 4.56″

Table 6: Antiprism dihedral angles.

A summary of the symmetry properties of prisms, antiprisms, and prismoids is given in Table 7 below. Similar tables will be found in succeeding chapters as appropriate. It should be noted that in Table 7, as will also be the case in Tables 9, 14, and 18, the symmetry groups and values of *n* are for a general member of the family. In many cases, there are individual members of the family with higher symmetry, such as the cube among the prisms, and in these cases, the values of *n* will be lower.

Name of polyhedron	Symmetry group	n-Isohedral	n-Isogonal	n-Isotoxal
		n	n	n
prism	D_{nh}	2	1	2
antiprism	S_{2nv}	2	1	2
prismoid	C_{nv}	3	2	3

Table 7: Primary symmetry properties of prisms, antiprisms, and prismoids.

Chapter 7. The Archimedean polyhedra.

It may be recalled that the Platonic solids were defined as polyhedra satisfying the following two criteria:

1. All the faces must be regular polygons of the same kind (*i. e.*, all triangles, all squares, etc.), and
2. All the vertices must have the same number of polygons meeting.

Suppose a slight relaxation of these: in number 1, the words "of the same kind" are omitted, but in consequence of this, the second should be revised to say "same number of polygons of any given kind," so if *any* vertex has, say, two hexagons and a triangle, *all* of them do. Such a polyhedron as can satisfy the two criteria as thus modified still has a great degree of regularity.

It should first, however, be noted that three equilateral triangles have a combined vertex angle total of 180°, as do two squares, and since it can be seen from Table 1 that *any* regular polygon has a vertex angle of less than 180°, it is clear that the requirement that the sum of the angles of polygons meeting at any vertex of a polyhedron be strictly less than 360° (see p. 18, Chapter 3) will always be satisfied by three equilateral triangles and *any* regular polygon, or by two squares and *any* regular polygon. If one builds up a polyhedron with three equilateral triangles and some regular polygon (the same type everywhere) at each vertex, it will be seen that an antiprism with equilateral triangular lateral faces results. And if one builds up a polyhedron with two squares and some regular polygon (the same type everywhere) at each vertex, it will be seen that a prism with square lateral faces results. Because these are *infinite* classes — the regular polygon that is additional to the three equilateral triangles or two squares can have any number of sides from 3 on up to infinity — these polyhedra are not included in the following discussion.

With the aid of Table 1, and omitting the prisms and antiprisms just mentioned, a limited group of polyhedra will be found whose faces are regular polygons and with the sum of the angles of polygons meeting at any vertex of a polyhedron being strictly less than 360°, obeying the other criteria given at the start of this chapter. These polyhedra are called the *Archimedean solids,* and although the prisms and antiprisms are not included in the normal definition of the Archimedean solids, some people have used the adjective "Archimedean" for prisms and antiprisms that obey these criteria as well. (The name "Archimedean" is applied because the Greek mathematician Archimedes is supposed to have discussed them, although the particular work of his in which he did is now lost. See the article "Archimedean solid" in *Wikipedia,* for example.) There are 13 different Archimedean solids, or 14 according to some definitions. (One definition states that there must be a symmetry transformation of the entire solid that takes one vertex to any other; in other words, the vertices must all fall into one transitivity class, or putting it yet another way, the polyhedron must be strictly isogonal. Another definition only requires that the faces meeting at each vertex form a congruent

Chapter 7. The Archimedean polyhedra.

configuration to that at any other vertex. The former definition is stricter, and a solid called the *pseudorhombicuboctahedron* or *elongated square gyrobicupola* satisfies the second but not the first. Norman Johnson did not consider the elongated square gyrobicupola to be an Archimedean solid, and so included it as J_{37} in his list of ninety-two.)

As the definition of the Archimedean solids concentrates on the polygons meeting at a vertex, the thirteen solids are best classified in that way. The *vertex figure* is a shorthand description of the types of polygons meeting at each vertex. If a vertex figure is given as "3·4·5·4" it means that there are a triangle, a square, a pentagon, and a square at each vertex. (It is given as 3·4·5·4 rather than 3·4·4·5, because the two squares are not adjacent; the pentagon is actually between the squares. A vertex figure of 3·4·5·4 could also, for example, be given as 5·4·3·4, but it is customary to start at the smallest number.) Table 8 below lists the thirteen Archimedean solids, arranged in order of their vertex figures. For each, the common name is given (in some cases, two names are both common), and in the columns headed F_3, F_4, etc., the total number of triangles, squares, etc. in the whole solid are given. The total numbers of all faces is shown in the column headed F_{total}, and the numbers of vertices and of edges are also given. (That the Euler formula, introduced on p. 20, Chapter 4, applies to all is easily verified.)

Vertex figure	Name	F_3	F_4	F_5	F_6	F_8	F_{10}	F_{total}	V	E
3·3·3·3·4	snub cube	32	6					38	24	60
3·3·3·3·5	snub dodecahedron	80		12				92	60	150
3·4·3·4	cuboctahedron	8	6					14	12	24
3·4·4·4	(small) rhombicuboctahedron	8	18					26	24	48
3·4·5·4	(small) rhombicosidodecahedron	20	30	12				62	60	120
3·5·3·5	icosidodecahedron	20		12				32	30	60
3·6·6	truncated tetrahedron	4			4			8	12	18
3·8·8	truncated cube	8				6		14	24	36
3·10·10	truncated dodecahedron	20					12	32	60	90
4·6·6	truncated octahedron		6		8			14	24	36
4·6·8	great rhombicuboctahedron (truncated cuboctahedron)		12		8	6		26	48	72
4·6·10	great rhombicosidodecahedron (truncated icosidodecahedron)		30		20		12	62	120	180
5·6·6	truncated icosahedron			12	20			32	60	90

Table 8: The thirteen Archimedean solids.

In Chapter 4, it was noted that the nomenclature of the Platonic solids is applied in a somewhat sloppy manner: terms like "tetrahedron" are used to denote "*regular* tetrahedron" although they can in truth be more generally applied. The same is true of some of the names listed in Table 8. In particular, all the names that contain the words "truncated" are here used in much more restricted ways than, in theory, they could be. Just as an example, the term "truncated cube" is here used to mean a figure like the one illustrated in Figure 20 below, where the six octagons are *regular*.

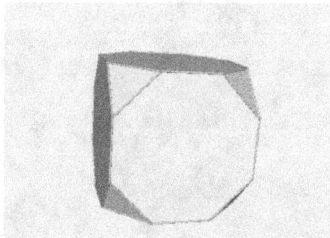

Figure 20: The Archimedean truncated cube.

Chapter 7. The Archimedean polyhedra.

In the same way, the term "truncated octahedron" here represents a figure like the one illustrated in Figure 21 below, where the eight hexagons are regular.

Figure 21: The Archimedean truncated octahedron.

In fact, the cuboctahedron (Figure 22) is just as much a "truncated cube" as the polyhedron illustrated in Figure 20, and also just as much a "truncated octahedron" as the one illustrated in Figure 21, although it is not generally called that, simply because it *is* both. In Chapter 9, below, the actual operation called "truncation" will be described in detail, and the full range of (not necessarily Archimedean) truncations of polyhedra will be considered. Therefore it should be understood that the restriction of such terms as "truncated cube" to Archimedean solids that can be so named is a matter of convention, and when such names are used, it should be verified whether the Archimedean solid or the more general case of a truncation of a named polygon is intended.

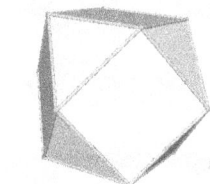

Figure 22: The cuboctahedron, an Archimedean solid that is both a truncation of a cube and the truncation of an octahedron.

The Archimedean solids, like the Platonic solids, have high degrees of symmetry. The truncated tetrahedron, like the tetrahedron among the Platonic solids, has T_d symmetry, while *five* Archimedean solids (the cuboctahedron, truncated cube, truncated octahedron, [small] rhombicuboctahedron, and great rhombicuboctahedron [truncated cuboctahedron]) all have the same O_h symmetry as the octahedron and cube among the Platonic solids (the snub cube has O symmetry, identical except for not having the mirror planes of O_h), and five Archimedean solids (the icosidodecahedron, truncated dodecahedron, truncated icosahedron, [small] rhombicosidodecahedron, and great rhombicosidodecahedron [truncated icosidodecahedron]) all have the same I_h symmetry as the icosahedron and dodecahedron among the Platonic solids (the snub dodecahedron has I symmetry, identical except for not having the mirror planes of I_h).

It is not only the requirement that the sum of the angles at any vertex must be less than 360° (see p. 18, Chapter 3) which must be obeyed. Since all vertices of an Archimedean solid have the *same* arrangement of polygons, and the angle defects (the *amounts* by which the sum of the angles at any vertex is less than 360°) must all sum to 720° (see p. 18, Chapter 3), each individual angle defect must be an *exact divisor* of 720°. This eliminates some combinations of polygons that might be expected, on the basis of having the sum of the angles at any vertex less than 360°, to be possible. For example, suppose one wanted to construct an Archimedean solid which has an equilateral triangle, a square, and a regular pentagon at each vertex (vertex figure 3·4·5). Consulting Table 1 on p. 15, one can see that these polygons have angles of 60°, 90°, and 108°, summing to 258°. This is considerably less than 360°, but this does not mean that such a polyhedron can be constructed. The angle defect is 102°, so seven such vertices would give a total defect of 714° (6° too small) and eight would give a total defect of 816° (96° too large). Thus no such Archimedean solid exists. (Of course, a 3·4·5 vertex could exist on some polyhedron together with *other types* of vertices, but such a polyhedron would not be Archimedean, because not all the vertices are alike.)

If one looks at Table 8 on p. 36, the angle defects can easily be computed for each of the 13 Archimedean solids. (For each *n*-gon, the defect is

$$180° - (n - 2)(180°)/n,$$

or one can simply refer to Table 1, on p. 15 in Chapter 2.) For example, the great rhombicosidodecahedron (truncated icosidodecahedron) has a vertex figure of 4·6·10, and the square, hexagon, and decagon respectively have vertex angles of 90°, 120°, and 144°, whose sum is 354° for an angle defect of 6°. There are 120 vertices, so the total defect is 720°.

The Archimedean solids, like the prisms and antiprisms, are isogonal. The number of transitivity classes to which the edges or faces belong is not the same for all thirteen Archimedean solids, but is tabulated in Table 9 below.

While there are other combinations of polygons that could seem to satisfy all these requirements (such as the vertex figure of 3·4·4·5 mentioned, and dismissed, a few

Chapter 7. The Archimedean polyhedra.

paragraphs earlier on p. 35), it becomes apparent that one cannot actually construct polyhedra satisfying the definition of an Archimedean solid with such combinations. Only the combinations listed in Table 8 (and the vertex figures 4·4·n, defining a prism, and 3·3·3·n, defining an antiprism) can be put together to form a complete polyhedron.

The reason for the name "rhombicuboctahedron" for two of these polyhedra is that twelve of the square faces (all twelve square faces of the great rhombicuboctahedron, 12 out of 18 of the small rhombicuboctahedron) determine planes which, if those planes alone are considered, would be the faces of a rhombic dodecahedron (see p. 22, Chapter 4). The two polyhedra named "rhombicosidodecahedron" have similar relationships with another figure, whose faces are all rhombi, named the the rhombic triacontahedron.

The great rhombicuboctahedron is sometimes called the *truncated cuboctahedron,* and the great rhombicosidodecahedron is sometimes called the *truncated icosidodecahedron,* but these names are misleading. If one applies the procedure named *truncation* (explained in Chapter 9) to a cuboctahedron or an icosidodecahedron, one obtains polyhedra closely resembling the great rhombicuboctahedron and the great rhombicosidodecahedron, but with rectangles at the faces where there ought to be squares. (See Figure 28 on page 48.) For this reason, the terms *truncated cuboctahedron* and *truncated icosidodecahedron* would appear to be incorrect, even though found in the literature, for the great rhombicuboctahedron and the great rhombicosidodecahedron.

It may be noted that the most symmetric of the prisms and antiprisms are Platonic solids (the cube, as a regular square prism, and the octahedron, as a regular triangular antiprism, both of symmetry O_h). By making the bases of a prism or antiprism to be regular polygons, it is always possible to make it "Archimedean" (though they are not considered among the Archimedean solids). The prisms of this type have symmetry D_{nh}, and the antiprisms S_{2nv} (=D_{nd}). Changing the heights (the distances between the planes containing the bases) reduces the symmetry of the *lateral faces* (they cease to be regular, becoming rectangles rather than squares in the case of the prisms, and isosceles, rather than equilateral, triangles in the case of the antiprisms), but the symmetry of the *polyhedron* is unchanged: D_{nh} or S_{2nv} as the case may be. As the bases are reduced in symmetry, the resulting prisms and antiprisms are also reduced in symmetry, but a *prism* retains a mirror plane in any case, so it can never have less than C_s symmetry. An *antiprism*, of course, is reduced to C_1 symmetry in the least symmetric case.

Normally, studies of polyhedra begin with the Platonic solids, treat the Archimedean solids next by relaxing conditions (as was done at the beginning of this chapter), and, by further relaxation of conditions, deal with further sets of polyhedra. For example, the great contribution of Norman Johnson was to treat the polyhedra obtained when the only condition set is that all the faces are regular polygons. It turns out that besides the Platonic and Archimedean solids, and the prisms and antiprisms, which have so far been discussed, there are ninety-two more polyhedra obtained, and while he did not prove that the ninety-two he found were the only ones in his 1966 paper, this was subsequently proved in 1969 by Victor Zalgaller. These ninety-two polyhedra are now designated the Johnson solids in his honor. Johnson gave these polyhedra names, and some of the nomenclature he invented is a basis for some terminology that will be used in this book. He also gave each one a number from 1 to 92, and in most treatments of these polyhedra, his numbers are still used, so that J_{57}, for example, is universally recognized as a designation for the fifty-seventh polyhedron in his list.

Another type of generalization one sees is the *uniform polyhedra*, whose definition follows the definition of the Archimedean polyhedra in this chapter but relaxes the restriction on polygons made in Chapter 2 in order to allow boundary-self-crossing polygons as faces (but still require the polygons to be *regular* in the sense of having all sides and vertices equal). Since we have made a conscious decision *not* to allow boundary-self-crossing polygons, nothing more will be said about the uniform polyhedra.

Instead of these generalizations, however, this book will take a different direction; going back to the prisms and antiprisms discussed in Chapter 5, we will deal with *infinite* families such as these — which will have some of the same properties that the families of prisms and antiprisms do, and which will, in fact, include some of the polyhedra that Johnson studied, but as single members of these infinite families. It is my belief that it is of more interest to look at polyhedra in this way than to consider small sets of polyhedra (granted that ninety-two, such as Johnson found, is not really a *small* number, but it is certainly small compared to infinity!) with very little that can be said about *all* members of a set.

A summary of the symmetry properties of the thirteen Archimedean solids is given in Table 9 below, in the same format as was used in Table 7.

Name of polyhedron	Symmetry group	n-Isohedral	n-Isogonal	n-Isotoxal
		n	n	n
snub cube	O	3	1	2
snub dodecahedron	I	3	1	2
cuboctahedron	O_h	2	1	1
(small) rhombicuboctahedron	O_h	3	1	2
(small) rhombicosidodecahedron	I_h	3	1	2
icosidodecahedron	I_h	2	1	1

Chapter 7. The Archimedean polyhedra.

Name of polyhedron	Symmetry group	n-Isohedral	n-Isogonal	n-Isotoxal
		n	n	n
truncated tetrahedron	T_d	2	1	2
truncated cube	O_h	2	1	2
truncated dodecahedron	I_h	2	1	2
truncated octahedron	O_h	2	1	2
great rhombicuboctahedron	O_h	3	1	3
great rhombicosidodecahedron	I_h	3	1	3
truncated icosahedron	I_h	2	1	2

Table 9: Primary symmetry properties of the Archimedean solids.

What all the polyhedra studied in the remainder of this book have in common is that they have a single axis of rotation of higher order than any other, which (as was mentioned on p. 8 in Chapter 1) is termed the *principal* axis of rotation (or principal rotation axis), and that they can be grouped into families which differ in the *order* of their principal rotation axis. Just as a prism has a number of mirror planes, twofold rotation axes, and such that is determined by the order of its principal rotation axis, these families will also have various additional symmetry elements, but the description of the family can be summarized by giving the symmetries as a function of this order. (For example, an *n*-gonal prism has an *n*-fold principal rotation axis, but it also has *n* + 1 mirror planes — including *n* passing through the axis and one perpendicular to it, and *n* twofold rotation axes other than the principal axis — of course, if *n* is even, the principal axis is a twofold axis as well!) Under certain circumstances, one of the members of such a family will be a Platonic or Archimedean solid, as in fact a cube is a type of prism, but because the definition of the family does not require such high symmetry, this will mean that one has treated one of the rotation axes as principal and ignored the others, just for the purpose of relating this high-symmetry polyhedron to a whole family of lower-symmetry, but related, polyhedra. (A cube has three fourfold axes, only one of which can be treated as principal, and four threefold axes, all of which must be ignored, when a cube is treated as a prism. It is actually possible, by treating *one of the threefold axes* as principal and ignoring all the *fourfold* axes, to put the cube into *another* family, which we will see on p. 81 in Chapter 21.)

While we have defined the principal rotation axis to be the highest-order rotation axis of a polyhedron, occasionally, as was just in the case of a cube in the preceding paragraph, it is useful to choose an axis that is equal in order to others (or even lower-order than one or more others) as a principal rotation axis in order to put a polyhedron into a family. And some of these families include, besides the principal rotation axis, additional twofold axes (as do the prisms and antiprisms), yet (unlike the prisms, but, as we will later find out, the antiprisms do belong in this category) the family can be said to include one member in which the principal rotation axis is twofold. When these cases seem naturally to belong to a family, it will be appropriate to allow the term "principal rotation axis" to be used in this way, even if it is *not* the highest-order rotation axis of the polyhedron in question.

However, before describing the various families, a series of chapters, beginning with Chapter 8, will follow in which some procedures that are useful to make new polyhedra (which will then be categorized into these families) are described.

Chapter 8. Modification procedures involving polyhedra 1. General considerations.

There are a number of modification procedures that can be applied to a polyhedron which change its nature, but produce a polyhedron that can be related to the original polyhedron. In fact, the mathematician John Horton Conway has devised a notation for describing polyhedra in terms of a number of modification procedures applied to one of the Platonic solids. This notation will not be used in this book, however, because it does not include some modification procedures that will be considered important for our discussion. (*Many* of the Conway procedures will be considered in this book, some will *not*, but we will consider *others* that Conway did not choose to treat.) If the reader is interested, the Conway notation can be found online at

http://www.georgehart.com/virtual-polyhedra/conway_notation.html

among other places. As stated there, nearly all the Conway procedures produce polyhedra with the same symmetry as the starting polyhedron, while this book will be heavily concerned with polyhedra whose symmetry is lower than the symmetries of the Platonic solids. In particular, the modification procedure that is described in this book as *uniaxial stretching* is of major importance.

Chapter 9. Modification procedures involving polyhedra II. Truncation and rectification.

One modification procedure that will be discussed at some length in this book is *truncation*. Suppose one starts with the cube illustrated in Figure 23 below. One can cut off a corner by a plane; however, the most useful process is to cut off all the corners in a manner that preserves the symmetry of the polyhedron. So in this case, one cuts it off in such a way as to intersect each of the original edges of the cube the same distance from the corner that was cut off.

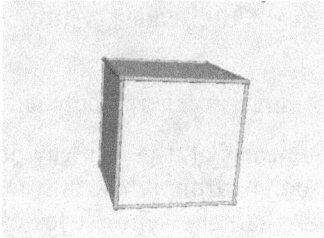

Figure 23: A cube, with no modification procedure applied.

This leads to a polyhedron such as the one illustrated in Figure 24 below. It should be noted that all three of the fourfold axes passing through the centers of the six original cube faces, all four of the threefold axes passing through the eight original cube vertices, and ll six of the twofold axes passing through the midpoints of the twelve original cube edges, remain. Each of the original vertices has been replaced by three, because three edges met at the original vertex. (If one started with, say, an icosahedron, where five edges meet at a vertex, each original vertex would become five.) The twelve original edges remain, though shortened, and twenty-four new edges are created. (In the original cube, there were eight vertices, each one located at the meeting point of three edges. In general, if one multiplies those two numbers — the *three* edges that meet at a vertex by the *eight* vertices truncated — the resulting product, *twenty-four,* will be the number of new edges made in the truncation.) The original six faces are changed from squares to octagons (but the octagons are not regular; they are, in the terminology of Chapter 2, 4×2-gons), and eight new faces are created, one for each original vertex. (In general, the *n*-gonal faces of the original polyhedron become 2*n*-gonal faces. As many new faces are created as the original polyhedron had vertices, and they will be polygons whose number of sides is the number of edges that meet at a vertex.)

Chapter 9. Modification procedures involving polyhedra II. Truncation and rectification.

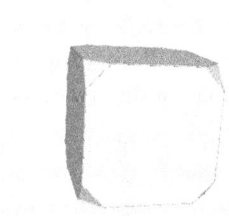

Figure 24: The cube, with a small degree of truncation.

As was mentioned above, the six octagons produced are not regular, but the sides of each alternate in length, with the sides that come from the original edges of the cube longer than the sides that are newly created by the truncation. This is why they qualify as 4×2-gons in the notation of Chapter 2. If one cuts further away from the original vertices, one finds a point where the two sets become equal, so the octagons become regular. If one starts off with a Platonic solid, at this point the resulting polyhedron, as shown in Figure 25, becomes Archimedean.

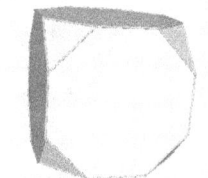

Figure 25: The Archimedean truncated cube.

If one carries the truncation further, one obtains a polyhedron like that illustrated in Figure 26. Once more, as in Figure 24, the octagons are not regular, but alternate in length, meaning that the polygons are again 4×2-gons; however, unlike the case of Figure 24, the sides that come from the original edges of the cube are *shorter* than the sides that are newly created by the truncation. As was noted in Chapter 7, when one is discussing the Archimedean solids, the term "truncated cube" normally means the polyhedron illustrated in Figure 25. However, the term "truncated cube" in general should refer to any of the polyhedra illustrated in Figures 24 through 26. In this book, this more general meaning will be used unless otherwise stated. If a sentence in this book refers to a "truncated ‹anything›," however, it will be assumed that all vertices are truncated as in the examples of Figures 24 through 26. If one were to truncate some but not all, an indication of how many vertices are involved will be used; for example, a "*tetratruncated cube*" is a figure derived by truncation of *four* of its eight vertices. (There are more ways than one, however, of doing this, so further specification of the truncation is necessary; subsequent chapters will go into this.) If an "Archimedean truncated ‹anything›" is intended, the word "Archimedean" will be explicit in general.

There is one exception commonly made to the nomenclature of the previous paragraph. Truncation of a pyramid at its apex produces a prismoid, which should also, by the terminology of that paragraph, be termed a "monotruncated pyramid." However, the normal term for this polyhedron is a "*frustum* of a pyramid." There is no common name for the specific truncation of a polyhedron which has a vertex located on the principal rotation axis (which will be seen, in Chapter 18 below, to be what in this book is termed an *apical polyhedron*) in such a way as to remove that vertex. Because the polyhedron produced thus from a pyramid is commonly termed a "*frustum* of a pyramid," however, it seems best to call such a modification procedure "*frustification.*" It will also be desirable to use the term "*frustum*" for any polyhedron produced by this kind of truncation.

Figure 26: The cube, with a greater degree of truncation.

Eventually, one reaches a point where the original edges of the polyhedron are reduced to zero. the only edges remaining are the twenty-four newly-created ones. The six faces, originally square, become square once more, but rotated 45° relative to their original position. (For any Platonic solid, the n-gonal faces, which were made $2n$-gonal through the truncation process, become once more n-gonal, but rotated by $180°/n$.) At this point, the truncation is referred to as a *rectification*. However, the rectified Platonic solids all have special names, so terms such as "rectified cube" are not in common usage. The cube and octahedron, when rectified, produce the same figure, which because of the fact that it is produced from both, is known as a *cuboctahedron*. This polyhedron was met before, among the Archimedean solids listed in Table 8 of Chapter 7. Figure 27 below illustrates this polyhedron.

Chapter 9. Modification procedures involving polyhedra II. Truncation and rectification.

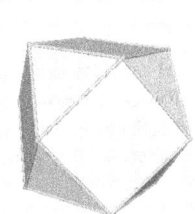

Figure 27: The cuboctahedron produced by rectifying the cube previously illustrated.

Similarly, rectification of either an icosahedron or a dodecahedron produces the figure listed in Table 8 as an *icosidodecahedron*, but while the cube, octahedron, icosahedron, and dodecahedron all yield Archimedean solids, the tetrahedron, on rectification, yields a Platonic solid: the octahedron.

If one cuts deeper, allowing the planes to be even further away from the vertices of the original cube, one creates new edges where the truncation planes intersect each other. The squares that were found in the cuboctahedron become smaller, while the triangles become hexagons, and the figure is recognizable as a truncated octahedron; eventually the truncation planes meet in sets of four (consisting of the four that had cut a single face of the original cube), the original faces of the cube become single points (vertices of a new polyhedron) and the polyhedron has become an octahedron. If we had started with an octahedron, we would by the same process have ended up with a cube when all three truncation planes cutting a face of the original octahedron become concurrent in a single point.

One can designate the stages as "shallow," "critical," and "deep" truncation. Shallow truncation refers to the process where some of the original edges attached to the truncated vertex remain as edges, now meeting the newly created face at a new vertex. In shallow truncation of a cube, a truncated cube, as was shown in the stages between the original cube and the cuboctahedron, is created. Critical truncation refers to the precise point when the truncation planes from the vertices at opposite ends of an edge meet at the midpoint of the edge, reducing it to zero. So critical truncation of a cube produces a cuboctahedron. Critical truncation, in this sense, is a synonym of rectification, and is only used when one is making this threefold distinction; we will continue to use the term "rectification" whenever we are not opposing it to shallow or deep truncation. And finally, deep truncation refers to the case where the truncating planes are further from the vertices than the critical position, causing new edges to be generated where they intersect each other, and producing a figure better thought of as a truncation of the dual than of the original polyhedron.

Of course, the cuboctahedron itself can be truncated, and if the truncation is at the point where the edges are all equal, the truncated cuboctahedron is itself an Archimedean solid. It can be called an Archimedean truncated cuboctahedron, but the name "great rhombicuboctahedron" has also been given to it. Truncated cuboctahedra, like the other truncated polyhedra discussed in this chapter, are not necessarily Archimedean, of course. One example of a truncated cuboctahedron is illustrated in Figure 28 below.

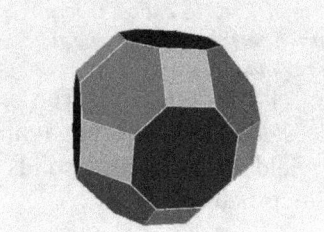

Figure 28: A truncated cuboctahedron, close to Archimedean, but not exactly so.

The truncated cuboctahedron illustrated in Figure 28 departs from Archimedean in that its edges are not all equal; if the average edge length is given as 1, there are 24 of the 72 edges whose lengths are approximately 1.233, 24 edges whose lengths are approximately 0.970, and 24 edges whose lengths are approximately 0.797 units. The symmetry is still O_h, as would be the symmetry of a true Archimedean truncated cuboctahedron, all vertices are 2.2412 units from the centroid of the polyhedron, all faces have the same angles they would have if regular (all 90° for the quadrilaterals, so they are rectangles, all 120° for the hexagons, and all 135° for the octagons).

One can conceive of a process, similar to truncation, where what is cut off are not the *vertices*, but the *edges* of a polyhedron. This is termed *cantellation*, although the term is generally applied only when *every* edge of the polyhedron is cut off in this way. Since each edge meets others at its endpoint, a large number of new faces are created: one for *each* edge of the original polyhedron *and* one for each vertex (where edges meet). For example, a cube becomes a small rhombicuboctahedron. Because this process leads to very complicated polyhedra even when starting with simple-to-describe ones, it will not be treated in detail in this book; it is only mentioned, here in this paragraph, and nothing further will be said of cantellation.

Chapter 10. Modification procedures involving polyhedra III. Augmentation.

Augmentation, in a sense, is the opposite of truncation. When a truncation occurs, an original vertex is eliminated, and a new face is created, sharing an edge with every one of the faces that had met at that vertex. A new set of vertices is added, one on each of those edges that had met at the eliminated vertex. In terms of the Euler formula $V + F = E - 2$ (see p. 20, Chapter 4), if n was the number of edges (or faces) meeting at the truncated vertex, V is increased by $n - 1$, F is increased by 1, and E is increased by n. In an augmentation, an original *face* is eliminated by conceptually "gluing" a small pyramid to it. If that face had been an n-gon, the n lateral faces of the pyramid now belong to the augmented polyhedron, and the augmented polyhedron also has n new edges, namely the ones that had met at the apex of the pyramid. A new vertex is created: the apex of the new pyramid. So V is increased by 1, F is increased by $n - 1$, and E is increased by n in terms of the Euler formula (see p. 20, Chapter 4).

The term "*cumulation*" is also used for this modification procedure, as, for example, in the online *Mathworld Encyclopedia*, where it is the preferred term, but "*augmentation*" is listed as an alternative term, as well as two others: "*accretion*" and "*akisation*." The latter comes from a nomenclature that has been adopted for fully augmented Platonic solids, where if every n-gonal face of the polyhedron is replaced by an n-gonal pyramid, the prefix *triakis, tetrakis*, or *pentakis* (for n = 3, 4, or 5) is used, as "*pentakis dodecahedron*." Perhaps the term "*akisation*" should be reserved for a *full* augmentation, namely, an augmentation of *every face* of a polyhedron. Outside of the *Mathworld Encyclopedia*, however, the only term I have ever seen is "*augmentation*," and so this will be the term used in this book.

While one has a lot of freedom in locating the cutting plane in a *truncation*, essentially being able to place it anywhere that intersects all the edges that had met at the eliminated vertex, one needs to take more care in an *augmentation*. If the new vertex is too far away from the face that was added to, the resulting polyhedron will be a nonconvex polyhedron, which is certainly not a desirable thing. Basically, if one looks at the n planes of the faces that shared an edge with the eliminated faces, and these planes, if extended, would all meet in a point, the new apex (which can be considered the apex of the pyramidal augmentation) must be closer to the plane of the face where the pyramid is "glued" onto than that meeting point. (If these n planes do not all intersect in a single point, it is necessary to look at the various points in which they *do* intersect and make sure that the new vertex is closer to the plane of the "gluing" than *all* of them.)

Occasionally, the term "augmentation" has been used in a broader sense. For example, the Johnson solid J_{66} has been termed an "augmented truncated cube," because it is generated by adding a polyhedron called a "square cupola" (see p. 119, Chap. 29) to one of the octagonal faces of an Archimedean truncated cube. This usage is not general, however, and in this book, it will be the case that the term "*augmentation*" will be reserved for the process described in this chapter, and *"fusion"* (See Chapter 14) will denote the more general operation of combining two different polyhedra in general.

While "augmented ‹anything›" often implies *full augmentation*, it will often be useful to speak of augmenting only one, two, or some other specific number of faces of a polyhedron, so terms like "monoaugmented," "biaugmented," etc., can be used. Just as was discussed for truncation (p. 46, Chapter 9), however, it may be necessary to specify which faces of a polyhedron are augmented.

Chapter 11. Modification procedures involving polyhedra IV. Uniaxial stretching.

Consider a cube, such as the one illustrated in Figure 29 below. It is possible to stretch it by keeping all the x- and y-coordinates unchanged, while multiplying all the z-coordinates by a constant, producing a rectangular prism such as the one illustrated in Figure 30.

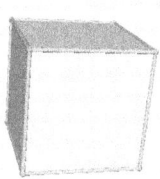

Figure 29: A cube, before any stretching.

Since the stretching is in only one direction, it is termed a *uniaxial* stretching of the cube; one could stretch the cube simultaneously along more than one axis, of course, but the uniaxial stretch is important for the purposes of this book, and any stretching along multiple axes can be considered as a sequence of uniaxial stretchings along different axes anyway, so only uniaxial stretchings will be considered here. In transforming the cube illustrated in Figure 29 to the rectangular prism illustrated in Figure 30, many of the properties are retained, but clearly the symmetry is *reduced*. For example, the fourfold symmetry around the z-axis remains, while the x- and y-axes are no longer fourfold axes. The threefold axes which pass through opposite vertices of the cube are all gone, but the twofold axes remain along each of the lines bisecting two opposite edges.

Figure 30: The prism obtained from the cube of the previous figure by a uniaxial stretching.

Because a cube is such a simple figure, it is probably easier to see some of the properties of a uniaxial stretching by using a more complex shape, such as the regular dodecahedron illustrated in Figure 31 below.

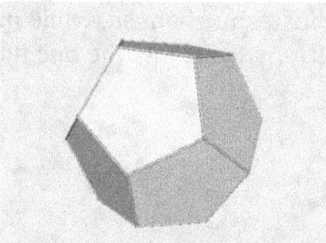

Figure 31: A dodecahedron, before any stretching.

The stretching of this dodecahedron along an axis through the midpoints of two opposite faces produces the polyhedron of Figure 32. The two pentagons through which the stretching axis passes remain regular; the other ten pentagons are now no longer *regular,* but only *isosceles* (or *strombic*; remember that an isosceles polygon of an odd number of sides is also strombic, as indicated on p. 14, Chapter 2, above.)

Again, many, but not all, of the properties of the regular dodecahedron remain. The fivefold axis (actually a tenfold alternating axis) through the centers of the base and roof remains; the five other fivefold axes are gone. The five twofold axes perpendicular to the stretching axis remain; but many of the twofold axes no longer exist. And while the original dodecahedron had ten threefold axes, none of those are found in the stretched dodecahedron.

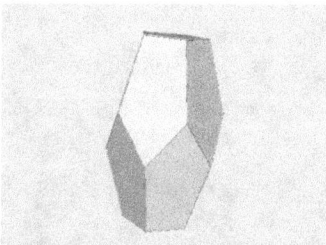

Figure 32: The dodecahedron from the previous figure, stretched along a fivefold axis.

Chapter 11. Modification procedures involving polyhedra IV. Uniaxial stretching.

It should be noted (and this can be seen by examining the two examples above of the cube and dodecahedron) that if the stretching is done along an axis which is a rotational symmetry axis of the original polyhedron, it will remain a rotational symmetry axis, of the same order, of the new, stretched polyhedron. Any mirror planes through that axis, as well as mirror planes perpendicular to that axis, will be preserved as well. Other rotational symmetry axes, in general, will *not* be preserved, though twofold rotational symmetry axes perpendicular to the axis of stretching will be preserved, and higher-order (only if *even*, however) rotational symmetry axes perpendicular to the axis of stretching will remain rotational symmetry axes, but reduced to twofold.

When, therefore, a Platonic or Archimedean solid undergoes a stretching along any of the rotational symmetry axes that it possesses, it still retains a rotational symmetry axis of the same order, even though regular polygons (other than those in planes perpendicular to the rotational symmetry axis) cease to be regular. Since many of the treatments of polyhedra that can be found restrict their discussion to regular-polygon-faced polyhedra, these uniaxially stretched Platonic or Archimedean solids are not considered worth discussion in those treatments. However, in this book, these and other polyhedra that contain rotational symmetry axes (termed *axially symmetric polyhedra*) will be considered in detail. The remainder of this book, in fact, will be devoted to a description and classification of axially symmetric polyhedra.

Chapter 12. Modification procedures involving polyhedra V. Dualization (general concepts).

In Chapter 4, the concept of duality of polyhedra was introduced. It was noticed that for each polyhedron listed in Table 2, there was another polyhedron in the table, termed the *dual* to it (or in the case of the tetrahedron, the *same* polyhedron had the appropriate properties), such that:

1. The number of *faces* of the original polyhedron was equal to the number of *vertices* of the dual polyhedron,
2. The number of *vertices* of the original polyhedron was equal to the number of *faces* of the dual polyhedron,
3. The number of *edges* of the original and dual polyhedra are equal,
4. The number of edges of the original polyhedron *bounding each face* was equal to the number of edges of the dual polyhedron *meeting at each vertex*, and
5. The number of edges of the original polyhedron *meeting at each vertex* was equal to the number of edges of the dual polyhedron *bounding each face*.

Looking at these statements, it should be noted that they pair up in sets of two (except for item 3. Since it speaks only of *edges*, there is no paired statement.) in which each reference to a *face* in one and each reference to a *vertex* in the other, and vice versa, correspond. So even these descriptive statements have a duality as significant as the duality of the polyhedra to which they refer.

In a general polyhedron, to define a dual it would be desirable to have the same properties, but in that case, the last two of these statements would often be meaningless, since "the number of edges meeting at each vertex" is not the same for each vertex, and "the number of edges of the original polyhedron bounding each face" is not the same for each face. So these statements have to be modified. In order to produce a proper modification, valid for all polyhedra, it is desirable even to replace the three statements at the beginning by somewhat differently-worded statements that would be equivalent to them, but which address each individual vertex, edge, and face, and again the statements will be paired in dual pairs:

1. To each *vertex* of the original polyhedron corresponds a *face* of the dual polyhedron,
2. To each *face* of the original polyhedron corresponds a *vertex* of the dual polyhedron,
3. The *vertices* at the corners of each *face* of the original polyhedron correspond individually to the *faces* meeting at the corresponding *vertex* of the dual polyhedron,
4. The *faces* meeting at each *vertex* of the original polyhedron correspond individually to the *vertices* at the corners of the corresponding *face* of the dual polyhedron,

Chapter 12. Modification procedures involving polyhedra V. Dualization (general concepts).

5. The number of edges *bounding each face* of the original polyhedron is equal to the number of edges *meeting at the corresponding vertex* of the dual polyhedron, and

6. The number of edges *meeting at each vertex* of the original polyhedron is equal to the number of edges *bounding the corresponding face* of the dual polyhedron.

7. To each *edge* of the original polyhedron corresponds an edge of the dual polyhedron such that:

   a) The two *faces* sharing that edge of the original polyhedron correspond to the two *vertices* at the endpoints of the corresponding edge of the dual polyhedron.

   b) The two *vertices* at the endpoints of that edge of the original polyhedron correspond to the two *faces* sharing the corresponding edge of the dual polyhedron.

It can be seen that this set of criteria provides more precision than the earlier set, because instead of simply stating the numbers of faces, edges, and vertices of the dual polyhedron, there is set up a specific correspondence between individual faces, edges, and vertices of the original polyhedron and vertices, edges, and faces of the dual polyhedron.

What remains is actually to produce the dual of any particular polyhedron. It should be noted that one only has to establish one of the three sets of elements (vertices, edges, or faces) of the dual polyhedron, because that set of correspondences determines the others, using the rules above. For example, if a *dual vertex* is assigned to each *original face*, rule 7.a) will fix the dual edges, and the dual faces will be fixed by rule 4.

The simplest modification procedure is, given a face, to select a point on it as the corresponding vertex of the dual. To preserve symmetry, whenever a face is a regular polygon, the dual vertex is chosen at the center of that polygon. When a face is not a regular polygon but has a rotocenter or mirror line(s), the dual vertex will be chosen at that rotocenter or on the mirror line(s). (If there is only one mirror line and no rotocenter, however, this does not fix the precise location of the dual vertex, but only constrains it to that line.) Then the edges of the dual polyhedron can be constructed by joining the vertices corresponding to the original faces that share edges, and this will similarly construct the dual faces.

This procedure is in fact related to the process of truncation described in Chapter 9. For if one starts with what was there called a deep truncation of any polyhedron at all its vertices, performed in such a way as to preserve as much symmetry as possible, the limiting (maximum depth possible) case of this truncation consists of moving the truncation planes as far from the vertices they are truncating as is necessary to bring the planes cutting each face to concur at a single point in the center of that face, and in that case, the truncation produces the dual as just described.

Although this modification procedure will actually be used on occasion in this book, it suffers from two problems: First of all, the resulting dual is much smaller than the original polyhedron. While it is desirable to have the dual of the dual be the original polyhedron, using this modification procedure on a tetrahedron produces a "dual of the dual" which is only one ninth the size of the original, as is shown in Appendix C. Similar reductions apply to all the Platonic solids. Secondly, a certain amount of arbitrariness will, in some cases, apply to the choice of a point to fix the dual vertex. The first of these objections is often made by first applying the modification procedure, and then enlarging the dual polyhedron to such a size that its edges intersect the corresponding edges of the original polyhedron. (In the case of the Platonic solids, this is always possible, and the midpoint of each dual edge will coincide with the midpoint of the corresponding original edge. In general, this will not always be possible, but in many cases it will be, or at least a large number of edges can be made to intersect their correlates.) But, except when symmetry is relatively great, such as in the case of the Archimedean solids, the second of these problems, the arbitrariness of the choice of at least *some of* the points to serve as dual vertices, remains. There is a remedy to this problem, but it is more complicated than can be clearly explained here, and it will therefore be reserved to the following chapter, which is optional.

In any case, however, a "conceptual dual" can be established. The exact locations of the elements might not be determined, but one can say that there are a certain number of triangular, square, ..., faces, that there are a certain number of vertices of order 3, 4, ..., that (for example) three triangles and a square meet at each of five vertices, and such. When it is not necessary to give precise measurements of the edge lengths, angles, etc., this conceptual dualization will often suffice.

Note that in the Euler formula (p. 20, Chapter 4), the numbers of faces and vertices appear symmetrically, so that it is clear that the Euler formula allows for dualization.

As it is desirable to preserve all symmetries of a polyhedron when constructing the dual (as was stated earlier), it can be seen that if the original polyhedron is n-isogonal, the dual will be n-isohedral (and vice versa) and if the original polyhedron is n-isotoxal, the dual will also be n-isotoxal. (Note that "isogonal" and "1-isogonal" mean the same thing, as do "isohedral" and "1-isohedral" and also "isotoxal" and "1-isotoxal.")

It should be noted that, through the concept of duality, the two modification

Chapter 12. Modification procedures involving polyhedra V. Dualization (general concepts).

procedures of truncation (Chapter 9) and augmentation (Chapter 10) are related in a special way: If a polyhedron is *augmented,* the *dual of the augmented polyhedron* is the same type of polyhedron as is obtained by *truncating the dual of the original polyhedron,* where the face that is augmented corresponds to the vertex of the dual that is truncated.

Chapter 13. Modification procedures involving polyhedra VI. Dualization (mathematical methods). *[Optional]*

In the previous chapter, some rough-and-ready, mainly intuitive methods for generating the dual of a specified polyhedron were given. For many polyhedra, particularly ones with mostly regular polygons for faces or with a high degree of symmetry, these methods work well. But a more general method, termed *polar reciprocation*, has the advantage that it will work in a more general case where the methods of the previous chapter will not.

In order to perform the polar reciprocation procedure, it is necessary to pick a center and a scale. When the polyhedron has an obvious center point, the most satisfactory dual is obtained by choosing that as the center of the polar reciprocation procedure; if not, it is still desirable to preserve any symmetries that exist in the polyhedron by choosing, as the center of the polar reciprocation, some point along any axis of symmetry that the polyhedron possesses. If there is no symmetry of the polyhedron to preserve, of course, the center of the polar reciprocation is arbitrary, but none of the polyhedra that will be discussed in this book will be totally asymmetric. The scale is chosen by specifying a polar reciprocation radius.

The formulas that will be presented here assume that the center of the polar reciprocation is the origin. Performing a polar reciprocation procedure with a center elsewhere is best accomplished by a three-step process:

1. Translating the polyhedron so that the point chosen as the center of the polar reciprocation is the origin of the new coordinate system,

2. Performing the polar reciprocation procedure as described below, with the origin as center of the procedure,

3. Translating the dual polyhedron generated in the previous step back to the original coordinate system.

The choice of a center will determine the result; the same polyhedron will yield very different results if different centers are used. But the choice of a polar reciprocation radius only determines the scale; changing the modification procedure radius will make the resulting polyhedron smaller or larger, but its shape will not be changed. The first step in the procedure is to generate a "polar plane" from each point. With the origin as the center of the procedure and a polar reciprocation radius r, the polar plane corresponding to a given point whose coordinates are (x_0, y_0, z_0) is the plane whose equation is given by

$$x_0 x + y_0 y + z_0 z = r^2.$$

It should be noted that if P is any point in a plane p', and the polar plane of P is a plane p while there is a point P' whose polar plane is p', then P' is a point in the plane p. For let the coordinates of P be (x_0, y_0, z_0), and the coordinates of P' be (x_0', y_0', z_0'). As just stated, the equation of p is $x_0 x + y_0 y + z_0 z = r^2$, while it is obvious that the equation of p' is $x_0' x + y_0' y$

Chapter 13. Modification procedures involving polyhedra VI. Dualization (mathematical methods).
[Optional]

$+ z_0'z = r^2$. But stating that P is a point in the plane p' implies that the coordinates of P satisfy the equation defining p', or in other words:

$$x_0'x_0 + y_0'y_0 + z_0'z_0 = r^2.$$

But this is exactly the condition required to specify that P' is a point in the plane p, since there is no difference between saying

$$x_0'x_0 + y_0'y_0 + z_0'z_0 = r^2$$

and

$$x_0x_0' + y_0y_0' + z_0z_0' = r^2.$$

Now let us consider a face of the original polyhedron which is a polygon whose vertices are $P_1, P_2, ..., P_n$. Let the polar planes of $P_1, P_2, ..., P_n$ be $p_1, p_2, ..., p_n$. Since all of $P_1, P_2, ..., P_n$ are points in one common plane, say p', the point P', whose polar plane is p', must simultaneously be in all the planes $p_1, p_2, ..., p_n$. But it has already been stated that $P_1, P_2, ..., P_n$ are the vertices of a face (located in plane p') of the original polyhedron, so portions of the planes $p_1, p_2, ..., p_n$ (which we can take to be faces of a new polyhedron) all meet at P', which we can make a vertex of the new polyhedron.

Let us review the definition of a dual polyhedron in the previous chapter:

1. To each *vertex* of the original polyhedron corresponds a *face* of the dual polyhedron,
2. To each *face* of the original polyhedron corresponds a *vertex* of the dual polyhedron,
3. The *vertices* at the corners of each *face* of the original polyhedron correspond individually to the *faces* meeting at the corresponding *vertex* of the dual polyhedron,
4. The *faces* meeting at each *vertex* of the original polyhedron correspond individually to the *vertices* at the corners of the corresponding *face* of the dual polyhedron,
5. The number of edges *bounding each face* of the original polyhedron is equal to the number of edges *meeting at the corresponding vertex* of the dual polyhedron, and
6. The number of edges *meeting at each vertex* of the original polyhedron is equal to the number of edges *bounding the corresponding face* of the dual polyhedron.
7. To each *edge* of the original polyhedron corresponds an edge of the dual poly-

hedron such that:

a) The two *faces* sharing that edge of the original polyhedron correspond to the two *vertices* at the endpoints of the corresponding edge of the dual polyhedron.

b) The two *vertices* at the endpoints of that edge of the original polyhedron correspond to the two *faces* sharing the corresponding edge of the dual polyhedron.

This process obviously satisfies the first four criteria, since the construction was based on exactly these criteria. It is not hard to verify that the remaining criteria are satisfied as well, so that this procedure will *always* construct a proper dual, *provided* that the faces are properly bounded (*i. e.*, that the edges bounding them do not run off to infinity) and that nothing crosses in the newly constructed polyhedron. This book will not attempt to establish the conditions for this to be true, but simply recommend that when polar reciprocation is used, the resulting figure should be checked to make certain it is a proper polyhedron.

Software for applying this procedure exists: the *Antiprism* package that I have used to help generate the figures in this book, among others, contains a program that can be used to generate dual polyhedra by polar reciprocation from polyhedra whose vertex coordinates and face connections are supplied.

Consider the following set of coordinates, which define a polyhedron which, in a later chapter, will be shown to be what will there be termed a *pentagonal 4-pyramoid*:

P_1: (0.0, 0.0, 2.0)

P_2: (1.0, 0.0, 1.0)

P_3: (0.309017, 0.951057, 1.0)

P_4: (-0.809017, 0.587785, 1.0)

P_5: (-0.809017, -0.587785, 1.0)

P_6: (0.309017, -0.951057, 1.0)

P_7: (1.309017, 0.951057, 0.0)

P_8: (-0.5, 1.538842, 0.0)

P_9: (-1.618034, 0.0, 0.0)

P_{10}: (-0.5, -1.538842, 0.0)

P_{11}: (1.309017, -0.951057, 0.0)

Chapter 13. Modification procedures involving polyhedra VI. Dualization (mathematical methods).
[Optional]

The faces are the following:

Five quadrilaterals: $P_1P_2P_6P_3$, $P_1P_3P_8P_4$, $P_1P_4P_9P_5$, $P_1P_5P_{10}P_6$, and $P_1P_6P_{11}P_2$,

Five triangles: $P_7P_3P_8$, $P_8P_4P_9$, $P_9P_5P_{10}$, $P_{10}P_6P_{11}$, and $P_{11}P_2P_7$,

and one pentagon: $P_7P_8P_9P_{10}P_{11}$.

A view of this polyhedron is given below in Figure 33:

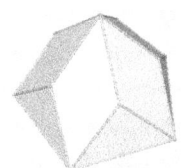

Figure 33: A pentagonal 4-pyramoid.

Applying the polar reciprocation procedure, one obtains a dual polyhedron specified as follows:

P_1: (0.681818, 0.495370, 1.318182)

P_2: (-0.260432, 0.801525, 1.318182)

P_3: (-0.842773, 0.0, 1.318182)

P_4: (-0.260432, -0.801525, 1.318182)

P_5: (0.681818, -0.495370, 1.318182)

P_6: (0.258285, 0.794921, 0.894649)

P_7: (-0.676201, 0.491288, 0.894649)

P_8: (-0.676201, -0.491288, 0.894649)

P_9: (0.258285, -0.794921, 0.894649)

P_{10}: (0.835830, 0.000000, 0.894649)

P_{11}: (0.0, 0.0, -0.824675)

The faces are the following:

One pentagon: $P_5P_1P_2P_3P_4$,

Five triangles: $P_1P_5P_{10}$, $P_2P_1P_6$, $P_3P_2P_7$, $P_4P_3P_8$, and $P_5P_4P_9$, and

Five quadrilaterals: $P_{11}P_6P_1P_{10}$, $P_{11}P_7P_2P_6$, $P_{11}P_8P_3P_7$, $P_{11}P_9P_4P_8$, and $P_9P_{11}P_{10}P_5$.

With a bit of inspection, it can be seen that this is, in fact, also a pentagonal 4-pyramoid, almost identical to the original but inverted (the apex of the dual, at P_{11}, is in the negative-z direction rather than the positive-z direction).

Chapter 14.   Modification procedures involving polyhedra VII. Fusion.

In Chapter 10, the term *augmentation* was introduced. It was noted there that, although defined differently, augmenting a polyhedron was equivalent to constructing a pyramid whose base was congruent to the face that was to be augmented, and conceptually "gluing" the small pyramid (at its base) to that face. In fact this is a special case of a process that can be described as *fusion*, namely, conceptually "gluing" two polyhedra together if each one has a face that is congruent to a face of the other. And it should be noted that if the faces in question are centered on rotational symmetry axes of the respective polyhedra, the resulting fused polyhedron will also be rotationally symmetric about that axis.

When it is proposed to build a polyhedron by fusing two together, it is important to consider the dihedral angles between the face being fused and every adjacent one, on both polyhedra. If the sum of the dihedral angle at a particular edge of one polyhedron and the dihedral angle at the edge that is being fused with it on the second polyhedron is 180° or greater, the fusion cannot be accomplished — at least, if it is greater, the resulting polyhedron will not be convex, and if it is equal, two faces will be in the same plane; in this case, one could perform a fusion, but these two faces will have to be combined into a single face of the new polyhedron. (The fact that the dihedral angles sum to *less* than 180° does not guarantee convexity of the resulting polyhedron, but if the dihedral angles sum to *more* than 180°, it certainly guarantees *non*convexity of the resulting polyhedron.)

It will be useful to use the term "*simple polyhedron*" to denote any polyhedron that cannot be obtained by fusing two polyhedra along a common face, and "*composite polyhedron*" to denote any polyhedron that *can* be obtained by fusing two polyhedra along a common face. Thus, a tetrahedron is a simple polyhedron, as is any other pyramid; the octahedron (as any bipyramid) and icosahedron (as any biaugmented antiprism) are composite polyhedra. Most of the polyhedra considered in this book will be simple, but since some of the Platonic and Archimedean solids are composite, it will be necessary to treat some composite polyhedra, especially those which are related to the Platonic and Archimedean solids in some way. Sometimes it will be useful to decompose a composite polyhedron into the separate polyhedra which were fused to produce it; this will be done in Chapter 15, for example.

Chapter 15. Modification procedures involving polyhedra VIII. Elongation and gyroelongation.

The terms *elongation* and *gyroelongation* were introduced by Norman Johnson with slightly more restricted meanings than will be used in this book. (The only significant difference comes from his insistence on considering only polyhedra with regular faces; thus his processes involve fusion with prisms whose lateral faces are squares and antiprisms whose lateral faces are equilateral triangles only.)

If a polyhedron has an *n*-fold principal axis of rotational symmetry, then it is necessary that for every vertex of the polyhedron, there are $n - 1$ others related to it by rotation (with one exception, which will be dealt with below), and these, together with the original vertex, all fall on a plane perpendicular to the rotational symmetry axis. Such a plane containing *n* polyhedron vertices arranged in a regular *n*-gon (sometimes, actually, some multiple of *n* polyhedron vertices arranged in a polygon containing a multiple of n vertices, but still *n*-fold symmetric!) will be called a *vertex rotation plane* of the polyhedron. The one exception is that a vertex that is *located on the rotational symmetry axis* will be solitary.

Consider a polyhedron in which all the vertices in some particular vertex rotation plane are connected by edges of the polyhedron. (If the polyhedron is simple, as defined on p. 63, Chapter 14, this is only possible if the polyhedron has a base; if it is composite, the plane where fusion had taken place is the vertex rotation plane to consider.) It may be possible to fuse a prism to this plane (if the original polyhedron was simple), or to decompose the polyhedron into two pieces and fuse those pieces to the bases of a prism (if the original polyhedron was composite). The only reason that the term "may" is used in this statement — one could *always conceive* of such a fusion taking place — is that the result of such a fusion will in many cases not be a convex polyhedron, and if that is the case, one would not like to perform such a fusion. This is *elongation*. The same process, using an *antiprism* rather than a *prism*, is called *gyroelongation*, because the S_{nv} symmetry of the antiprism requires that, when the original polyhedron has to be decomposed into two pieces prior to the fusion, one piece must be rotated 180°/*n* around an axis perpendicular to the vertex rotation plane in question.

As stated earlier, when Johnson proposed the terms *elongation* and *gyroelongation*, he envisioned the prism having only squares for lateral faces (and regular polygons for bases, with the original polyhedron having a regular polygon where the prism is fused), and the antiprism having only equilateral triangles for lateral faces, with regular-polygon bases, similarly. However, some writers, at least, have used these terms without this restriction. (For example, as I write these words, the *Wikipedia* article entitled "Gyroelongated bipyramid" refers to gyroelongated bipyramids that require isosceles, rather than equilateral, triangles on the antiprism fused into the combination; *e. g.*, the gyroelongated hexagonal bipyramid). In this book, Johnson's restrictions are dropped and this more general definition is accepted.

Chapter 16. Modification procedures involving polyhedra IX. Parallel cropping.

The procedure that will be introduced in this chapter under the name of "parallel cropping" (but will be simply referred to as "cropping" in subsequent chapters, as we will have no need to consider any other sort of cropping, except under a different name) is one which, to my knowledge, is not considered elsewhere, probably because, when applied to most regular-polygon-sided figures, the regularity of some of the faces is destroyed.

The procedure resembles truncation, as defined in Chapter 9. Just as in the truncation procedure, a plane is passed through the polyhedron, and everything on one side of that plane is discarded. The difference from truncation is that, in truncation, *one vertex* is discarded and the plane cuts all edges that meet at that vertex; in parallel cropping, the plane is passed parallel to a face, and the whole face is discarded; the plane cuts all edges that terminate at any vertex of that face. For the purposes of this book, since our focus is on axially symmetric polyhedra, we will always crop faces in planes perpendicular to the principal rotation axis, which will preserve all the rotations around that axis as symmetry transformations of the polyhedron.

Parallel cropping is not always a productive procedure. When applied to a prism, it simply has the same effect as a uniaxial stretching of the prism. When applied to a prismoid, it just produces another prismoid, and when applied to a pyramid, it produces a smaller copy of the same pyramid. Where it is most important is when it is applied to certain other polyhedra, and specifically we will be discussing cropping of *antiprisms*, because the resulting figures will be referred to in later chapters.

Consider the hexagonal antiprism shown in Figure 34 below. Cropping of just one of the two bases destroys the S_{12v} symmetry; the twelvefold alternating axis is replaced by a simple sixfold rotation axis and the monocropped hexagonal antiprism now has only C_{6v} symmetry. Half of the triangles along the sides are preserved as isosceles triangles; the others become isosceles trapezoids. Cropping of the other base as well, however, can restore the S_{12v} symmetry if the cropping is done the same distance from the original base, and the remaining triangles become trapezoids as well in the resulting bicropped hexagonal antiprism. Though both the original antiprism and a symmetrically bicropped one have the same symmetry – S_{12v}, or as some might prefer, D_{6d} – one looks in vain for any reference to bicropped antiprisms, by any name. (In this book, the term "quasiprism" will mean a symmetrically bicropped antiprism, because it has a number of properties resembling a prism, which will be explained later. It should be understood, however, that an antiprism can also be cropped asymmetrically on the two bases, by cropping different distances from the original planes of those bases.) Most books on polyhedra prefer to consider only those that can be formed of regular polygonal faces, and while, if the ratio of height to diameter is fixed correctly, a regular hexagonal antiprism can be built with equilateral triangles as lateral faces, the quasiprism cannot be given squares as lateral faces — the lateral faces *must* be isosceles trapezoids — and this lessens their interest for those other writers.

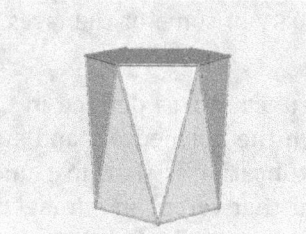

Figure 34: A regular hexagonal antiprism.

A view of the monocropped hexagonal antiprism produced from the antiprism of Figure 34 is given in Figure 35. The alternation of triangles and isosceles trapezoids as lateral faces is obvious from the figure.

Figure 35: A monocropped hexagonal antiprism, derived from the previously illustrated object.

The bicropped hexagonal antiprism produced by a symmetrical cropping of both bases is illustrated in Figure 36 below. All the properties mentioned earlier can be seen to apply to it.

Chapter 16. Modification procedures involving polyhedra IX. Parallel cropping.

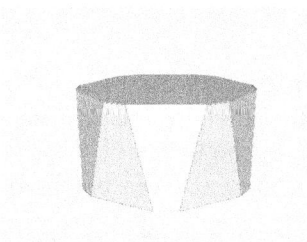

Figure 36: A symmetrically bi-cropped hexagonal antiprism, produced from the previous object by a second parallel cropping.

Chapter 17. Modification procedures involving polyhedra X. Convergence.

The term "convergence" is used in this book for a modification procedure that has not, to my knowledge, been discussed elsewhere (just as the cropping in Chapter 16 has not been), but perhaps should be considered in the light of the fact that our focus is on axially symmetric polyhedra, rather than on regular-polygon-faced polyhedra as is frequent.

Consider an axially symmetric polyhedron. On the principal rotation axis, but *outside* the polyhedron, select a point, to be designated the *convergence point*. For each vertex in the polyhedron, scale the vertex rotation plane that it is in, in proportion to the distance of that plane from the convergence point. This means that all edges parallel to the principal rotation axis are transformed into *converging* lines (hence the name) all concurrent in the convergence point. It can be seen that selecting the convergence point *outside* the polyhedron is important; if it were on the surface of the polyhedron, all points in the vertex rotation plane through the convergence point would be converted to a single point coincident with the convergence point. If this is on a face, it would convert the face to a vertex and eliminate the edges bounding that face. (While there have been other modification procedures, such as truncation and augmentation, that changed the numbers of faces, edges, and/or vertices of a polyhedron, it would not be a good thing to use the same name, in this case *convergence*, for a procedure which *did* change these numbers as for a procedure which did *not*.) A convergence point *inside* the polyhedron would give a "polyhedron" whose edges crossed each other, and we have agreed not to deal with such pathological cases.

Specifically, this procedure converts a *prism* to a *prismoid*. In most cases, no names exist for the new polyhedra thus created, so terms like "*converged antiprism*" will be used, as is the case for augmentation and truncation. (For an example of a converged antiprism, see Figure 55 on p. 110.)

Chapter 18. Classification of axially symmetric polyhedra.

Axially symmetric polyhedra will be the main focus of the remainder of this book. Because of that, it will be useful to introduce some terminology that will be used to classify these polyhedra.

Earlier, the concept of vertex rotation plane was defined. It was remarked that for every vertex of the polyhedron, there are $n - 1$ others related to it by rotation (with one exception, which will be dealt with below), and these, together with the original vertex, all fall on a plane perpendicular to the rotational symmetry axis. Thus, n polyhedron vertices would be arranged in a regular n-gon (sometimes, actually, some multiple of n polyhedron vertices will be arranged in a polygon containing a multiple of n vertices, but still n-fold symmetric!) in a plane termed a *vertex rotation plane* of the polyhedron.

The one exception, mentioned above, is that a vertex that is located on the rotational symmetry axis will be solitary. If a coordinate system is set up with the z-axis along the rotational symmetry axis of the polyhedron, such a vertex must be either the highest-z or the lowest-z point of all the vertices. If it is the highest-z point, it will be termed the *apex* of the polyhedron; if the lowest-z point, the *antiapex* of the polyhedron. A polyhedron containing an apex will be called an *apical* polyhedron; a polyhedron containing an apex and an antiapex will be called a *diapical* polyhedron. (One will never need to deal with a polyhedron containing an antiapex but not an apex; in that case, just reverse the direction of the z-axis and it is simply an apical polyhedron.) When it is desired to refer to an apical polyhedron that is not diapical, it will be called an *apicobasal* polyhedron to emphasize that it has an apex and a base opposite that apex. A polyhedron with no apex, having all its vertices in vertex rotation planes, will be called a *tectal* polyhedron (from the Latin word for *roof*). It is not necessary to distinguish polyhedra with a "roof" alone from one with a "roof" and a "base." Tectal polyhedra *by definition* have a base, since they would otherwise, by reversing the z-axis, qualify as apicobasal. Thus every axially symmetric polyhedron is either diapical, apicobasal, or tectal.

It will be convenient to consider the vertex rotation planes of a polyhedron as including one through the apex, if it is apical; and if the polyhedron is diapical, another vertex rotation plane through the antiapex will be assumed as well. It will often be useful to number the vertex rotation planes: the *first* vertex rotation plane will be the one through the apex (if the polyhedron is apical) or through the roof polygon (if the polyhedron is tectal), and the second, third, etc. will be the ones encountered as one travels away from the apex or roof polygon. When coordinates are assigned to the vertices (especially in the optional chapters where they are worked out in detail), while in some cases they will simply be numbered $P_1, P_2, ...$, it will often be the case that a lettering will be employed to take account of the vertex rotation plane in which the vertex is located: A (or $A_1, A_2, ...$) for points in the first vertex rotation plane; $B_1, B_2, ...$ for points in the second vertex rotation plane; etc. This lettering will be described in this book as "*vertex rotation plane based labeling.*"

The principal classification of polyhedra in this book will first be as diapical, apicobasal, or tectal. Apical polyhedra will secondly be classified by the order of the principal axis of rotational symmetry and third, by the polygons immediately adjacent to the apex. Tectal polyhedra will secondly be classified by the relationship between the number of sides of the roof and base polygons.

It should be noted that highly symmetric polyhedra such as the Platonic and Archimedean solids have multiple axes of rotational symmetry, and can be classified as diapical, apicobasal, or tectal with reference to each type of axis (but will be in the same category with respect to all axes of the same type). So a cube is *tectal* when considered in terms of one of its *fourfold* axes as a principal rotational axis, but *diapical* when considered in terms of one of its *threefold* axes as a principal rotational axis. As will be seen later, this is because it falls into two different polyhedron types (actually three, but only two are being considered here!): a *(tectal) prism* and a *(diapical) antibipyramid*. When a polyhedron is uniaxially stretched along the axis used to define it in such a category, it remains in that category, so that selecting to orient a cube so that its principal axis is along a threefold axis, in which case it is being treated as an antibipyramid, and stretching it along that axis, yields polyhedra which are also antibipyramids.

In Chapter 9, it was mentioned that in some cases, it will be necessary to specify a subset of vertices of a polyhedron to be truncated. One term that I believe is new in this book will come from this; if all the vertices in a single vertex rotation plane, but no others, are truncated, the term *peritruncated* (from the Greek prefix meaning "around") will be used. For this purpose, vertex rotation planes consisting of a single vertex (an apex or antiapex) of an apical polyhedron are not considered, and if the polyhedron is tectal, the base and roof are not usually considered as well. If this still does not specify the vertex rotation plane, because there are two or more in the polyhedron other than apex, antiapex, base, or roof), it will still require more specification, but it will usually be clear.

It should be obvious, considering the definition of duality in Chapter 12, that because the dualization procedure converts vertices to faces and vice versa, the dual of a *tectal* polyhedron is a *diapical* polyhedron and vice versa, while the dual of an *apicobasal* polyhedron is another apicobasal polyhedron, oriented the opposite way (that is, the apex of the dual points away from the apex of the original polyhedron), often, but not always, the same type of polyhedron.

In the subsequent chapters, the different sorts of axially symmetric polyhedra will be treated, arranged according to the classification scheme elucidated in this chapter. The Platonic and Archimedean solids, when they fit into the category of a particular chapter, will be mentioned in that chapter.

Chapter 19. Apical polyhedra with triangles at the apex 1. General considerations.

If triangles are placed with their vertices at the apex of a polyhedron, the simplest arrangement possible is to make them isosceles (or equilateral; this introduces no additional symmetry except in the presence of other symmetry elements, as in the Platonic solids). In that case, the edges meeting at the apex are all equal, and the sides of these triangles which are opposite to the apical vertex are all in one plane, forming a regular polygon in that plane. If the polyhedron is provided with a face constituting this regular polygon, it will form a base, and the construction of that polyhedron is completed. This polyhedron is, of course, a pyramid, the simplest apical polyhedron with triangles at the apex. A pyramid can be glued onto the roof of any *tectal polyhedron*, if that roof is congruent to the base of that pyramid, forming a new type of polyhedron which is apicobasal; as well, any *apicobasal polyhedron* whose base is congruent to the base of some pyramid can be inverted in its orientation (so the base becomes, in effect, a roof), and the pyramid glued onto that face, forming a new diapical polyhedron. Thus (since a pyramid is apicobasal), two pyramids with the same type and size of base can be glued together, forming a diapical polyhedron called a *dipyramid* or *bipyramid*; it can be seen that no apical polyhedron with triangles meeting at the apex can be anything other than a pyramid or such a fusion of a pyramid onto the roof of another polyhedron. (Such a fusion is termed *augmentation*, and was discussed in Chapter 10, although it is really a special case of fusing two polyhedra which have a face that is congruent together, and as such, perhaps should not be considered a fundamental modification procedure.) One also has to take a little care, as just randomly fusing a pyramid onto a polyhedron may lead to a result which is not convex, though by uniaxial stretching (really *compressing*, since the "stretch" decreases the height of the pyramid) the pyramid, it can always be changed to a form that will not introduce a departure from convexity.

Among the Platonic solids, the regular tetrahedron is a pyramid with a triangular base, and the regular octahedron is a square bipyramid. The regular icosahedron is actually a pentagonal antiprism which has been fused with pyramids on both the roof and base (formally, a *biaugmented pentagonal antiprism*, or a *gyroelongated pentagonal bipyramid*). Thus the antiprism (which is a tectal polyhedron) is converted to an apicobasal polyhedron, a *gyroelongated pentagonal pyramid* or *augmented pentagonal antiprism*, by the first fusion, and this intermediate polyhedron to a diapical polyhedron, the icosahedron, by the second.

As was mentioned earlier, Norman Johnson studied polyhedra whose faces were all regular polygons. In Chapter 3, it was noted that the vertex angles of all the faces coming together at a vertex must have a sum which is *strictly less* than 360°. The angles of an equilateral triangle are each 60° (See Table 1), so no more than *five* equilateral triangles can meet at the apex of a pyramid. Thus he dealt only with $n = 4$ and $n = 5$ in his list, where n denotes the number of triangles meeting at the apex, or, what is the same thing, the number of sides of the base polygon. (The case $n = 3$, of course, is a Platonic solid, the regular tetrahedron.) The square pyramid is J_1 and the pentagonal pyramid is J_2 in Johnson's list. In general, of course, *any* pyramid can be fused with an inverted copy of itself to form a bipyramid, but only if the base is triangular, square, or pentagonal will it be possible to use this process to form a polyhedron whose faces are all regular. A square bipyramid whose faces are all regular, as was stated in the previous paragraph, is a regular octahedron (a Platonic solid), while the triangular and pentagonal bipyramids with all faces regular are Johnson solids, designated as J_{12} and J_{13} in Norman Johnson's list. Similarly, *any* regular prism or antiprism can be fused with a pyramid, as was done in the first step of creating an icosahedron in the previous paragraph. Fusion with a prism produces an elongated pyramid or augmented prism (both names are equally appropriate); fusion with an antiprism produces a gyroelongated pyramid or augmented antiprism (both names are equally appropriate). An elongated pyramid can be created with all its triangles regular, of course, only if its base has fewer than 6 sides; the elongated triangular, square, and pentagonal pyramids are J_7, J_8, and J_9 in Norman Johnson's list. It should be noted that a *gyroelongated pyramid* can be created with all its triangles regular *only* if the base is square or pentagonal. Trying to do it with a triangular base will produce a figure in which the pyramidal sides and the antiprism sides adjacent to them are in the same plane, which means that what is actually produced is that which will in Chapter 21 be termed a *4-pyramoid*. The square and pentagonal gyroelongated pyramids with all faces regular are Johnson solids, designated as J_{10} and J_{11} in Norman Johnson's list. And the same condition, that the number of triangles meeting at the apex is no more than 5, applies of course to elongated and gyroelongated *bipyramids*. All these are summarized in Table 10 below.

For the cases where two alternative names can be given, namely the elongated and gyroelongated pyramids and bipyramids, which can also be thought of as mono- and biaugmented prisms and antiprisms, only the set of names is given in Table 10 that was used by Johnson, for space reasons; however, it must be understood that in each case, both names are equally appropriate and there is no reason to prefer the name "gyroelongated bipyramid" to "biaugmented antiprism."

It should be noted that when *no* requirement is made that all faces of a polyhedron be regular polygons, any number of triangles may meet at the apex, and the limitation of Table 10 to three columns is purely due to this limitation to polyhedra whose faces are all regular polygons. A gyroelongated hexagonal bipyramid (which would be the same as a biaugmented hexagonal antiprism) is certainly a possible polyhedron; it simply cannot be made with all the triangles equilateral. And thus, all six rows of Table 10 are really samples of *infinite* families of polyhedra, only a partial example of what was discussed on p. 40 in Chapter 7. And many other examples of apical polyhedra with triangles meeting at the apex are possible, but all are obtainable by augmentation of some other polyhedron. The only reasons for specifically listing the poly-

Chapter 19. Apical polyhedra with triangles at the apex I. General considerations.

hedra of Table 10 are because (firstly) they include three of the Platonic solids, and (secondly) they include many polyhedra met with very commonly in the literature (not only in Johnson's work).

Polyhedron	Type of base	Triangular	Square	Pentagonal
Pyramid		Platonic solid (regular tetrahedron)	Johnson solid (J_1)	Johnson solid (J_2)
Bipyramid		Johnson solid (J_{12})	Platonic solid (regular octahedron)	Johnson solid (J_{13})
Elongated pyramid		Johnson solid (J_7)	Johnson solid (J_8)	Johnson solid (J_9)
Gyroelongated pyramid		Not possible	Johnson solid (J_{10})	Johnson solid (J_{11})
Elongated bipyramid		Johnson solid (J_{14})	Johnson solid (J_{15})	Johnson solid (J_{16})
Gyroelongated bipyramid		Not possible	Johnson solid (J_{17})	Platonic solid (regular icosahedron)

Table 10: Regular-polygon-faced apical polyhedra with triangles at the apex.

It should be noted that *any* pyramid with a given type of regular polygon as a base can be converted to *any other* pyramid with the same type of regular polygon as base by uniaxial stretching, (at least in shape; a further scaling may be necessary, in general.) So in a sense *all* pyramids with the same type of regular polygon as base are related, and this book considers them so.

One point that should be emphasized is that the two pyramids that are fused together to form a bipyramid need not have the same heights, but when they do, the resulting bipyramid has D_{nv} symmetry, while when they do not, it has only C_{nv} symmetry. Similar comments apply to the elongated and gyroelongated bipyramids (biaugmented prisms and antiprisms) except that in the case of the biaugmented antiprism (gyroelongated bipyramid), the symmetry is S_{2nv} rather than D_{nv} if the pyramids have equal height. Only the bipyramids with D_{4v} symmetry qualify as "stretched octahedra." Similarly, not every biaugmented pentagonal antiprism will qualify as a "stretched icosahedron": to do so it must have both pyramids equal in height, and the ratio of the common height of the pyramids to the height of the antiprism must be the same as that of the icosahedron (which is computed in Appendix G). As long as both pyramids are equal in height, the biaugmented pentagonal antiprism will have the requisite S_{10v} symmetry, but unless the height bears the proper ratio to that of the antiprism, it will not truly be possible to consider it the result of a uniaxial stretching of an icosahedron on the fivefold axis.

Chapter 20. Apical polyhedra with triangles at the apex II. Mathematical details. *[Optional]*

Although any polygon can serve as the base of a pyramid, in this chapter the emphasis is placed on pyramids with bases that are regular n-gons. When this is so, the faces (other than the base) are isosceles triangles (possibly equilateral, but even an equilateral triangle can be considered as isosceles by ignoring the fact that the edge along the base is equal to the others). With that decision made, and maximal symmetry assumed, the most logical coordinate system to use has the z-axis passing through the apex and the center of the base, and using the vertex rotation plane based labeling for points described on p. 69, Chapter 18, the coordinates of the vertices are given by:

A_1: $(0, 0, z_A)$,

B_1-B_n: $(r \cos[k-1]\alpha, r \sin[k-1]\alpha, z_B)$, where $k = 1, 2, ..., n$; $\alpha = 360°/n = 2\pi/n$ radians.

There are two types of edges: AB_k and $B_k B_{k+1}$. Let the edges AB_k and $B_k B_{k+1}$ be denoted by e and s respectively. It is easy to see that for all k, the edge $e = \sqrt{[r^2 + (z_A - z_B)^2]}$ since $\cos^2 \alpha + \sin^2 \alpha = 1$. The $B_k B_{k+1}$ edges (all equal to s) require more manipulations to compute, but it can be seen that from symmetry it is only necessary to compute $B_1 B_2$, and the square of this edge is equal to:

$$r^2[(1 - \cos \alpha)^2 + \sin^2 \alpha] =$$
$$r^2[(1 - 2\cos \alpha + \cos^2 \alpha) + \sin^2 \alpha] =$$
$$r^2(1 - 2\cos \alpha + 1) =$$
$$r^2(2 - 2\cos \alpha) =$$
$$2r^2(1 - \cos \alpha).$$

It can be noted that the only way that the edges are equal (thus making the lateral faces equilateral triangles) is if

$$r^2 + (z_A - z_B)^2 = 2r^2(1 - \cos \alpha), \text{ or}$$
$$(z_A - z_B)^2 = r^2(1 - 2\cos \alpha).$$

Since the left-hand side of the equation is positive, and so is r^2, this is only possible if $\cos \alpha < \frac{1}{2}$, which requires $\alpha > 60° = \pi/3$ radians, and in turn this requires $n < 6$. For this reason, Norman Johnson, who only considered polyhedra with regular polygonal faces, dealt only with $n = 4$ and $n = 5$ in his list. (The case $n = 3$, of course, is a Platonic solid, the regular tetrahedron.)

The triangles $AB_k B_{k+1}$, as stated, are isosceles with $AB_k = AB_{k+1} = e$, and the angles $B_k B_{k+1} A$ and $B_{k+1} B_k A$ are therefore equal. Because the sum of the angles of a triangle is always $180° = \pi$

Chapter 20. Apical polyhedra with triangles at the apex II. Mathematical details. [Optional]

radians, if the angle B_kAB_{k+1} is designated θ, the angles $B_kB_{k+1}A$ and $B_{k+1}B_kA$ are both equal to $\frac{1}{2}(180° − \theta) = \frac{1}{2}(\pi − \theta)$ in radians. The requirement that the sum of all the angles at a vertex be less than 360° (or 2π radians) implies that, in any pyramid such as is being specified, $\theta < \alpha$. In turn, the base angles $\frac{1}{2}(180° − \theta) = \frac{1}{2}(\pi − \theta) > \frac{1}{2}(180° − \alpha) = \frac{1}{2}(\pi − \alpha)$.

If one designates the midpoint of the edge B_kB_{k+1} by M_k, the triangle AB_kM_k is a right triangle. The line AM_k (an altitude of triangle AB_kB_{k+1}) is termed the *slant height* of the pyramid. (In the symmetrical pyramids we are discussing, it is the same for all k.) We have already determined that $AB_k = e = \sqrt{[r^2 + (z_A − z_B)^2]}$, so the slant height is given by

$$l = e \sin \tfrac{1}{2}(\pi − \theta) = \sqrt{[r^2 + (z_A − z_B)^2]} \sin \tfrac{1}{2}(\pi − \theta).$$

The segment B_kM_k is given by

$$B_kM_k = AB_k \cos \tfrac{1}{2}(\pi − \theta) = \sqrt{[r^2 + (z_A − z_B)^2]} \cos \tfrac{1}{2}(\pi − \theta),$$

but since the edge $s = B_kB_{k+1} = 2B_kM_k$ has already been determined to be equal to $2r^2(1 − \cos \alpha)$, one has

$$r^2(1 − \cos \alpha) = \sqrt{[r^2 + (z_A − z_B)^2]} \cos \tfrac{1}{2}(\pi − \theta).$$

From this equation, given any three of r, α (or n), $z_A − z_B$, and θ, the remaining one can be computed. And as a result, from any three of them, in addition, the edges e and s, the slant height l, and all other significant measurements of the pyramid can be determined as well. (It will normally be the case that the three quantities supplied are r, n, and $z_A − z_B$.)

Table 11 below gives some limiting values of some of these quantities (along with actual values of those that do not vary with $z_A − z_B$). In the table, the limiting values are the ones that would obtain if $z_A = z_B$, and in an actual pyramid, the values are either greater than or less than the tabulated quantities. The limiting value of e/r is not tabulated, as it is 1 in all cases.

In an actual pyramid, the values of θ and s/e are always less than the tabulated quantity, while the values of l/r and l/e are always greater than the tabulated quantity. Since $e/r>1$ in all cases, in fact $l/r>l/e$, with both being greater than the tabulated quantity.

n	Actual value of α		Actual value of s/r	Limiting value of l/r
	Limiting value of θ		Limiting value of s/e	Limiting value of l/e
	degrees	radians		
3	120.000000	2.094395	1.732051	0.500000
4	90.000000	1.570796	1.414214	0.707107
5	72.000000	1.256637	1.175571	0.809017
6	60.000000	1.047198	1.000000	0.866025
7	51.428571	0.897598	0.867767	0.900969
8	45.000000	0.785398	0.765367	0.923880
9	40.000000	0.698132	0.684040	0.939693
10	36.000000	0.628319	0.618034	0.951057
11	32.727273	0.571199	0.563465	0.959493
12	30.000000	0.523599	0.517638	0.965926

Table 11: *Some quantities associated with regular-polygon-based pyramids.*

For example, suppose the square pyramid with coordinates

A: (0, 0, 1),

B_1: (1, 1, 0),

B_2: (−1, 1, 0),

B_3: (−1, −1, 0),

B_4: (1, −1, 0).

This is, in fact, rotated 45° with respect to the coordinate system that has just been assumed, but with that understood, this is a pyramid with $r=\sqrt{2}$, $n=4$, $z_A=1$, and $z_B=0$. It can be calculated rather easily that $l=\sqrt{2}$, $e=\sqrt{3}$, and $s=2$; the value of θ is harder to calculate but, by using the dot product formula of Appendix B, one finds $\cos\theta = 1/3$, implying that $\theta \approx 70.53°$. One can see that, comparing these quantities with those tabulated in Table 11, $\theta \approx 70.53° < 90°$, $s/r=\sqrt{2}=1.414214$, $s/e=2\sqrt{3}/3 \approx 1.1547 < 1.414214$, $l/r=1>0.707107$, and $l/e=\sqrt{6}/3 \approx 0.8165 > 0.707107$.

It might be noted that the requirement that the value of s/e is always less than the quantity tabulated in the column of Table 11 confirms the statement, made earlier on p. 74, that a pyramid with only regular polygons for faces (which requires equilateral triangles for lateral faces) is only possible if the base is a regular polygon of 3, 4, or 5 sides. For, if $n=6$ in Table 11, the tabulated limiting value of s/e is 1, so the actual value of 1 for an equilateral triangle, which must be *less than* the tabulated limiting value of s/e,

Chapter 20. Apical polyhedra with triangles at the apex II. Mathematical details. [Optional]

is impossible.

Chapter 21. Apical polyhedra with quadrilaterals at the apex 1. General considerations.

Four possibilities exist for apical polyhedra which involve only quadrilaterals meeting at the apex. In each case, following the edges of those quadrilaterals from the apex to the other end of that edge, we arrive at a vertex that cannot be a base vertex, because there will be yet another vertex of the quadrilateral further away from the apex. In other words, the fact that the polygons meeting at the apex are quadrilaterals implies that there are at least two more vertex rotation planes, one (here designated the *second* vertex rotation plane, in accordance with the numbering of the vertex rotation planes described on p. 69, Chapter 18) through the vertices of the quadrilaterals that are adjacent to the apical vertex, one (the *third* vertex rotation plane) through the remaining quadrilateral vertices (opposite the apical vertex in each quadrilateral). So there must be a new set of polygons with their nearest vertex to the polyhedral apex in that second vertex rotation plane. These might be triangles, so that the third vertex rotation plane is a base, beyond which there is nothing further from the apex; or they might be quadrilaterals, and in that case the second and third vertex rotation planes would be identical except for being inverted and rotated half a step. This second case is, obviously, diapical, with the fourth vertex rotation plane consisting of an antiapex alone, just as the first consists of an apex alone.

Yet a third case exists, similar to the second in that quadrilaterals complete the polyhedron, meeting at an antiapex, but unlike the second case described above, the second and third vertex rotation planes are not the same size. This case resembles the second, but the symmetry is different because turning the second type of polyhedron upside down and rotating it is a symmetry transformation, but for the third type it is not.

Finally, a fourth case exists, in which the edges that would, if extended to the axis, meet to form the antiapex are terminated short of that point, making these lateral polygons isosceles pentagons, and producing a *base* for the fourth vertex rotation plane. This case, like the first, is apicobasal.

In the first case, the polyhedron has C_{nv} symmetry (where n is the number of edges meeting at the apex, equal to the number of quadrilaterals meeting at the apex), much like a pyramid, and in fact is a pyramid with all vertices except the apex truncated (as regards the base, in fact, *rectified*). In this second case, the whole polyhedron would have S_{2nv} symmetry. The third and fourth cases, like the first, have C_{nv} symmetry.

There is no common name for the polyhedron described in the first of these three cases, but as it resembles a pyramid in so many particulars, in this book a new term will be introduced in this book: a *pyramoid*. Specifically, this is a *4-pyramoid*; others will be seen later. The following is the definition that will be used in this book.

Suppose an apicobasal polyhedron exists in which a number of strombic polygons (see

Chapter 21. Apical polyhedra with quadrilaterals at the apex I. General considerations.

p. 14, Chapter 2) all of the same type, meet at the apex, with edges from there extending to a second vertex rotation plane. In each of the remaining vertex rotation planes (if any), an edge simply runs from the vertex of one of the strombic polygons to the vertex of the adjacent strombic polygons. These edges define additional lateral faces (which will, if they exist, be isosceles triangles or trapezoids), and the final set of edges will form a regular polygon as a base. If the strombic polygons meeting at the apex are q-gons, the resulting polyhedron is an *q-pyramoid*. If there are k of them, it may also be specifically referred to as a *k-gonal q-pyramoid*.

By this definition, a pyramid is a 3-pyramoid, and that inclusive definition will be followed in this book. A 4-pyramoid can be obtained, as well as by the construction described in this definition, by starting with a pyramid and truncating it at every vertex except the apex, to the point that the truncation planes meet, so that every edge along the base is reduced to a single point (a rectification, if one only looks at those base vertices and ignores the apex and the edges that meet at the apex of the pyramid). If the truncation does not cut off as much, so that some of the base edges remain, the result of such a truncation is a 5-pyramoid.

A pentagonal 4-pyramoid is illustrated in Figure 37 below.

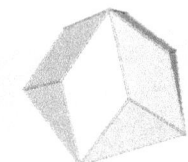

Getting back to the four kinds of polyhedron that could be generated from quadrilaterals meeting at an apex, the second, by contrast with the pyramid, has been given a plethora of names: *deltohedron, trapezohedron,* and *antibipyramid* among them. Since *deltohedron* is so similar to *deltahedron* (a name for a polyhedron formed of triangles only), it seems a bad name for this figure (though, because a kite-quadrilateral is sometimes referred to as a *deltoid,* the origin is clear). The name "trapezohedron" (although it appears to be the most frequently encountered name in the literature) would suggest a polyhedron bounded by *trapezoids,* which this is not, and in addition, the word is also used for at least one other type of polyhedron. So the preferred name in this book is "*antibipyramid.*" (This name, in fact, has one very good reason for it: An antibipyramid is the dual of an antiprism, as a bipyramid is the dual of a prism.)

The third case has not been considered anywhere of which I am aware. The names used for the second could be applied to this one, though the symmetry is different and it would be better to use a new name. Since all the faces are kite-quadrilaterals, in this book it will simply be referred to as a *polykite*, with it understood that an antibipyramid can be considered as a symmetrical polykite.

As for the fourth case, with isosceles pentagons meeting at a base, this polyhedron is basically an antibipyramid (or a polykite) with the far apex truncated. Because a polyhedron obtained by truncation of an apex alone is termed a *frustum* in the case of pyramids and bipyramids, we could extend this term and term this polyhedron a *monofrustum of an antibipyramid* (or *monofrustum of a polykite*). (It is a *monofrustum* because only *one* of the two apexes of the antibipyramid or polykite is truncated.) But this is a cumbersome term, and I prefer to coin a new one. If this shape is inverted, with base at the top and apex below, it resembles a faceted gem such as is mounted in a ring. The Greek *polytimos* means a gemstone, so I will call this figure a *polytimoid*. (Because the only real difference between an antibipyramid and a polykite is the symmetry — S_{2nv} for an antibipyramid, C_{nv} for a polykite — and the truncation of the apex destroys that additional symmetry of an antibipyramid, there is no particular reason to distinguish a polytimoid formed by truncating an antibipyramid from one formed by truncating a polykite, and no special names will be coined to distinguish one from the other.)

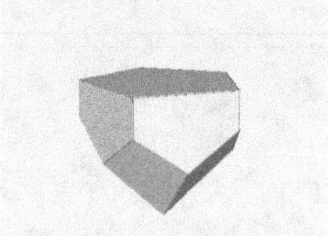

Figure 38: An example of a polytimoid, shown with the apex downward.

Figure 38 shows an example of a polytimoid, in this case a pentagonal polytimoid. In order to show the "gemstone-like" appearance of this polyhedron, which led to the coinage of the name, it has been shown with the apex pointing downward and the base at the top.

The same considerations that led to the conclusion that the only Platonic solid with square faces is the cube (p. 19) determine that no apical polyhedron exists with squares at the apex unless the principal rotation axis is *threefold*. It can be determined (see the next Chapter) that if two squares are placed, each with a vertex at the apex of the polyhedron and sharing an edge, the sides of these squares which are adjacent to the shared edge must make a right angle to each other. So if these squares were used to construct a 4-pyramoid, and those sides became edges of the polyhedron which are shared by each square with a triangle, this triangle must be a right triangle, and cannot be equilateral. Thus an investigation such as Norman Johnson's, in which only polyhedra with faces that are all regular polygons are considered, will not find the 4-pyramoids. And similarly, if the face beginning at the second vertex rotation plane is to be a pentagon, the requirement that this angle be a right angle makes it impossible for the pentagon to be regular.

A similar attempt to try to construct an antibipyramid whose sides are all regular, however, *succeeds* — and this antibipyramid turns out to be a *cube*.

The dual of a *4-pyramoid* is *another* 4-pyramoid, although the apex points in the opposite direction as the apex of the original 4-pyramoid. (See Figure 33, and the discussion regarding the dualization of that particular 4-pyramoid on page 61.)

The dual of an *antibipyramid* is an antiprism; as noted above, a cube is a regular triangular antibipyramid, so its dual is a regular triangular antiprism, which in fact is a regular octahedron. The dual of a polykite is a converged antiprism.

The dual of a *polytimoid* is a composite polyhedron which can be thought of, if the appropriate vertex rotation planes are the same size, as an augmented antiprism or a gyroelongated pyramid, a figure that was already encountered in Chapter 19, while if they are not it is best thought of as an augmented converged antiprism; in any case, it is obtainable by fusing a pyramid and an antiprism or converged antiprism, if the base of the pyramid is congruent to the bases of the antiprism, or to one base of the converged antiprism. An example of this figure is shown in Figure 39 below, which is the dual of the same polytimoid shown in Figure 38.

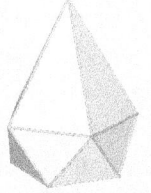

Figure 39: The dual of the polytimoid illustrated earlier in this chapter.

Chapter 22. Apical polyhedra with quadrilaterals at the apex II. Mathematical details. *[Optional]*

It will be useful to place the z-axis, as is usual in this book, along the principal rotation axis of the polyhedron. Vertex rotation plane based labeling is used to designate the vertices: the apex A will be assigned the coordinates $(0, 0, z_A)$, where z_A will be, as yet, arbitrary. Let the order of the principal rotation axis be n; then the points B_k, where $k = 1, 2, ..., n$, in the second vertex rotation plane, are given the coordinates $(r_B \cos \{k - 1\}\alpha, r_B \sin \{k - 1\}\alpha, z_B)$, where α is equal to $360°/n$, or $2\pi/n$ in radians. Again, both r_B and z_B can be considered arbitrary, although according to our convention for the z-coordinate, where the apex is the highest-z vertex, $z_B < z_A$. In order to have C_{nv} symmetry, the points C_k, where $k = 1, 2, ..., n$, in the third vertex rotation plane, must be located at the coordinates $(r_C \cos \{k - ½\}\alpha, r_C \sin \{k - ½\}\alpha, z_C)$, but this alone is not enough to guarantee that the points A, B_k, B_{k+1}, and C_k are all in one plane. However, if the values of r_B and r_C are chosen so that A, B_1, B_2, and C_1 are all in one plane, symmetry will assure that the remainder will be.

The easiest way to determine the location of C_1 is to use vectors (see Appendix B) and the trigonometric formulas of Appendix A. The vectors AB_1 and AB_2 can be computed as

$$AB_1 = (r_B, 0, z_B - z_A) \text{ and}$$

$$AB_2 = (r_B \cos \alpha, r_B \sin \alpha, z_B - z_A).$$

Vertex C_1, in order to be in the same plane as the three already located, needs to have the vector $AC_1 = m_1 AB_1 + m_2 AB_2$; but for symmetry reasons, $m_1 = m_2$, and we can simply write $AC = mAB_1 + mAB_2$, or $AC = m(AB_1 + AB_2) = (mr_B \{1 + \cos \alpha\}, mr_B \sin \alpha, 2m\{z_B - z_A\})$. This, in turn, requires:

$$r_C \cos \alpha/2 = mr_B (1 + \cos \alpha),$$
$$r_C \sin \alpha/2 = mr_B \sin \alpha,$$
$$z_C - z_A = 2m(z_B - z_A).$$

But $\cos \alpha = 2 \cos^2 \alpha/2 - 1$, so substituting this in the first of the three above equations leads to

$$r_C \cos \alpha/2 = mr_B (1 + \cos \alpha)$$
$$= 2 mr_B \cos^2 \alpha/2, \text{ or}$$
$$r_C = 2 mr_B \cos \alpha/2.$$

This automatically ensures that the second of the three equations is satisfied, because $\sin \alpha = 2 \sin \alpha/2 \cos \alpha/2$. This leaves the third, and that can be rearranged to give:

$$z_C = z_A + 2m(z_B - z_A)$$
$$= (1 - 2m)z_A + 2mz_B.$$

Now m can be chosen arbitrarily (though if $m<½$, $AB_1C_1B_2$ will not be convex, and if $m=½$, it becomes a triangle, so it is necessary to insist that $m>½$). For any value of $m>½$, then, setting $r_C = 2 mr_B \cos \alpha/2$ and $z_C = (1 - 2m)z_A + 2mz_B$ will make acceptable quadrilaterals for

Chapter 22. Apical polyhedra with quadrilaterals at the apex II. Mathematical details. [Optional]

$AB_1C_1B_2$.

If edges of the polyhedron are constructed along C_1C_2, C_2C_3, ... C_nC_1, then, a 4-pyramoid is produced.

If $2m \cos \alpha/2 = 1$, point C_1 will be exactly the same distance from the z-axis as points B_1 and B_2: this means that if a point D is located at $(0, 0, z_D)$ such that
$$z_D - \{(1 - 2m)z_A + 2mz_B\} = z_B - z_A, \text{ or}$$
$$z_D = \{(1 - 2m)z_A + 2mz_B\} + z_B - z_A$$
$$= (1 + 2m)z_B - 2mz_A,$$
the kite-quadrilateral $DC_1B_2C_2$ is congruent to $AB_1C_1B_2$. By constructing the edges C_1D, C_2D, C_3D, ..., C_nD, a polyhedron of S_{nv} symmetry is created, an *antibipyramid*.

The 4-pyramoid can also be constructed by considering it in a different way. If one starts with a pyramid and truncates all the vertices except the apex, doing it in a symmetric way with each truncation plane cutting the base at the two mirror planes that pass through two edge midpoints, the figure obtained is a 4-pyramoid of the same nature as the one that has been described. For this purpose, the vertices A, B_k, and C_k will be defined before, but a set of original vertices of the pyramid need to be fixed. They will be given coordinates
$$P_k: (r_P \cos \{k - 1\}\alpha, r_P \sin \{k - 1\}\alpha, z_C),$$
where it is already noted that the z-coordinate will be the same as that of all the C_k. It will be necessary to locate the points B_k and C_k, where each B_k is along the line AP_k and each C_k at the midpoint of the line segment P_kP_{k+1}. Furthermore, all the distances AB_k are equal.

Because of symmetry, again, one only needs to look at the points A, B_1, B_2, and C_1, and again, the formulas of Appendix B are useful. The coordinates of A, P_1, and P_2 are:
$$A: (0, 0, z_A),$$
$$P_1: (r_P, 0, z_C),$$
$$P_2: (r_P \cos \alpha, r_P \sin \alpha, z_C),$$

and the vectors AP_1 and AP_2 are:
$$AP_1 = (r_P, 0, z_C - z_A) \text{ and}$$
$$AP_2 = (r_P \cos \alpha, r_P \sin \alpha, z_C - z_A).$$
The value of z_B, as before, can be fixed arbitrarily, as long as $z_A > z_B > z_C$. Let
$$f = (z_A - z_B)/(z_A - z_C).$$
Then
$$AB_1 = fAP_1 = (fr_P, 0, f[z_C - z_A]) \text{ and}$$
$$AB_2 = fAP_2 = (fr_P \cos \alpha, fr_P \sin \alpha, f[z_C - z_A]),$$
but, since $f = (z_A - z_B)/(z_A - z_C)$, one can replace $f(z_C - z_A)$ by $z_B - z_A$, so
$$AB_1 = (fr_P, 0, z_B - z_A) \text{ and}$$
$$AB_2 = (fr_P \cos \alpha, fr_P \sin \alpha, z_B - z_A),$$
and the coordinates of B_1 and B_2 are

$$B_1 = (fr_P, 0, z_B) \text{ and}$$
$$B_2 = (fr_P \cos \alpha, fr_P \sin \alpha, z_B),$$

where $f = (z_A - z_B)/(z_A - z_C)$, so if one wants to write these coordinates in the form $B_k: (r_B \cos [k - 1]\alpha, r_B \sin [k - 1]\alpha, z_B)$ it is necessary to have:

$$r_P = r_B/f$$
$$= (z_A - z_C)r_B/(z_A - z_B).$$

With this value for r_P, and the fact that C_1 is at the midpoint of the line segment P_1P_2, we are ready to fix the coordinates of C_1. The x-coordinate is given by

$$r_P(1 + \cos \alpha)/2 =$$
$$(z_A - z_C)r_B(1 + \cos \alpha)/2(z_A - z_B),$$

and the y-coordinate is given by

$$(r_P \sin \alpha)/2 =$$
$$[r_B(z_A - z_C)\sin \alpha]/2(z_A - z_B).$$

Again, one would prefer to write these coordinates in the form $C_k: (r_C \cos [k - \frac{1}{2}]\alpha, r_C \sin [k - \frac{1}{2}]\alpha, z_C)$ so it is necessary to have:

$$[r_B(z_A - z_C)\sin \alpha]/2(z_A - z_B) = r_C \sin \alpha/2,$$

and writing $\sin \alpha = 2 \sin \alpha/2 \cos \alpha/2$, this becomes:

$$[r_B(z_A - z_C)\sin \alpha/2 \cos \alpha/2]/(z_A - z_B) = r_C \sin \alpha/2,$$

or, dividing by $\sin \alpha/2$,

$$[r_B(z_A - z_C)\cos \alpha/2]/(z_A - z_B) = r_C.$$

(We have not used the equation for the y-coordinate only because the one for the x-coordinate equation is simpler; if you wish to try it, you will receive exactly the same result.) With this equation, five quantities can be picked essentially arbitrarily: z_A, z_B, z_C, r_B, and n (since $\alpha = 360°/n = 2\pi/n$ radians), and they are sufficient to fix r_C. (They are not totally arbitrary, because, for example, it has been assumed that $z_A > z_B > z_C$.) In the first construction we found

$$r_C = 2 mr_B \cos \alpha/2 \text{ and}$$
$$z_C = (1 - 2m)z_A + 2mz_B.$$

For these equations, the five quantities z_A, z_B, r_B, m, and n (since, again, $\alpha = 360°/n = 2\pi/n$ radians) are sufficient to fix r_C and z_C. To show that the results are equivalent to the one just derived, it is only necessary to eliminate m from these equations:

$$r_C = 2 mr_B \cos \alpha/2,$$
$$m = r_C/(2r_B \cos \alpha/2), \text{ so}$$
$$z_C = (1 - 2m)z_A + 2mz_B$$
$$= z_A - 2m(z_A - z_B)$$
$$= z_A - r_C(z_A - z_B)/(r_B \cos \alpha/2),$$

or more conveniently,

$$z_A - z_C = r_C(z_A - z_B)/(r_B \cos \alpha/2).$$

But this can be seen to be identical to the equation just derived,

Chapter 22. Apical polyhedra with quadrilaterals at the apex II. Mathematical details. [Optional]

$$[r_B(z_A - z_C)\cos \alpha/2]/(z_A - z_B) = r_C,$$

with just a rearrangement of the factors, so both derivations yield the same coordinates for the vertices of the 4-pyramoid.

Chapter 23. Apical polyhedra with pentagons at the apex 1. General considerations.

For this chapter, it will be assumed that a polyhedron with at least C_{nv} symmetry is desired, and thus the pentagons we are discussing as meeting at the apex must be *isosceles*. If a set of *n* isosceles pentagons are fitted around the apex with their symmetry axes meeting there, two sides of each pentagon will form edges of the polyhedron that also meet at the apex, but since each edge belongs to two of these pentagons, the total number of edges meeting at the apex will be equal to *n*. Since we have assumed at least C_{nv} symmetry, all these n edges will be equal in length, and the bases of the pentagons will all lie in one plane. If a polyhedron is completed by constructing an edge joining each of the vertices at the base of each isosceles pentagon to the nearest vertex at the base of the adjacent isosceles pentagon, a set of *n* triangles and a 2*n*-gon are produced as faces. (The 2*n*-gon will not, in general, be regular, though by proper choice of parameters it can be made so; but it will be an *n*×2-gon according to the nomenclature of Chapter 2.) This polyhedron will be a *5-pyramoid*, as defined in Chapter 21. An example of a 5-pyramoid is shown in Figure 40 below.

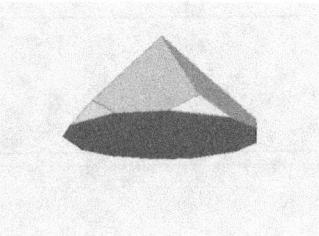

Figure 40: A right hexagonal 5-pyramoid.

The caption of Figure 40 identifies the object depicted as a *right hexagonal* 5-pyramoid because two specific kinds of information must be supplied to specify the details of any 5-pyramoid. First of all, it can be noted that the dihedral angles between the pentagonal faces and the base are always acute; they can never be otherwise. But the triangular faces may make acute, right, or obtuse dihedral angles with the base. And based on that, the 5-pyramoid is characterized as acute, right, or obtuse. Besides this, the value of *n* as defined above should be specified. Note that the base, being an *n*×2-gon, does not match the terminology; a *hexagonal* 5-pyramoid has a *dodecagonal* base.

As stated, the object in Figure 40 was constructed so that the triangles and the base have a 90° dihedral angle between them, so it is classified as a *right hexagonal* 5-pyramoid. An *acute hexagonal* 5-pyramoid is illustrated in Figure 41 below. It was actually constructed using all the vertices at the same points as in Figure 40 except for the base vertices, which were located closer to the points where the edges meeting at the apex would meet the base plane if extended.

Figure 41: An acute hexagonal 5-pyramoid, resembling the previous one.

If, instead, the base vertices were located *further* from the points where the edges meeting at the apex would meet the base plane if extended, an *obtuse* hexagonal 5-pyramoid, as shown in Figure 42 below, would be produced.

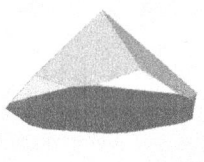

Figure 42: An obtuse hexagonal 5-pyramoid, resembling the previous one.

Because the base of a 5-pyramoid is an $n\times2$-gon, there are two conceivable ways that two 5-pyramoids can be fused together: by fitting them so the triangles of one share edges with the pentagons of the other, or by putting triangle to triangle and pentagon to pentagon. Using terminology developed by Norman Johnson for similar fused polyhedra that he considered, the first would use the prefix *gyro-* and the second the prefix *ortho-*. So we can call the first polyhedron type a *gyrobi-5-pyramoid* and the second a *orthobi-5-pyramoid*; however, the "5" can be omitted, because 3-pyramoids are pyramids, and both they and 4-pyramoids have bases that are only *n*-gons, so that the alternative fittings in the fusion are not possible. Thus it will be understood that the terms *gyrobipyramoid* and *orthobipyramoid* refer to fusions of 5-pyramoids only. In order to produce convex polyhedra, of course, it is necessary to be certain that the dihedral angles at the base are small enough. As was stated in Chapter 14, the sum of the dihedral angles at the edges of the base where fusion takes place must be less than 180°, or else the bipyramoid will not be convex. The *pentagonal* lateral faces will always form acute dihedral angles with the base, so it is only the *triangular* faces that need to be worried about. By choosing the parameters properly, however, the dihedral angles can be made small enough.

If one wants to fuse two identical 5-pyramoids to form an orthobipyramoid, they must be acute, and the resulting polyhedron is as illustrated in Figure 43 below. An orthobipyramoid can be made by fusing an acute and an obtuse 5-pyramoid, however, if the dihedral angle between the triangles in the acute 5-pyramoid and the base is small enough that, when combined with the dihedral angle between the triangles in the obtuse 5-pyramoid and the base, the sum is less than 180°.

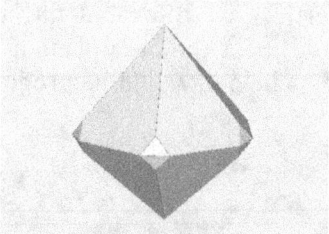

Figure 43: A symmetrical hexagonal orthobipyramoid.

Trying to fuse two identical *right* 5-pyramoids to form an orthobipyramoid is not possible, because the triangular faces of the two 5-pyramoids will be in the same planes. If, however, the edges separating each triangular face of the upper 5-pyramoid from the corresponding face of the lower are eliminated, so the two triangles merge to form a rhombus, a polyhedron is produced which is simple (no longer divisible into two along the plane that had been the bases) and which can be seen to be a bipyramid which has been truncated at all nonapical vertices. This polyhedron is illustrated in Figure 44 below. This polyhedron has not been named anywhere, but we have now provided enough nomenclature to name it: a *peritruncated bipyramid*.

Chapter 23. Apical polyhedra with pentagons at the apex I. General considerations.

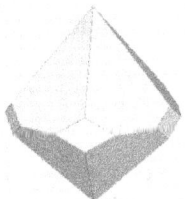

Figure 44: The peritruncated bipyramid obtained by fusing two right hexagonal 5-pyramoids and merging triangles into rhombi.

Because the $n\times2$-gon at the base of a typical 5-pyramoid is not a regular $2n$-gon, if one wishes to form an *gyrobipyramoid* by fusing it with another 5-pyramoid, the other 5-pyramoid is usually going to have quite different angles, and the resulting symmetry will be C_{nv}, just as each of the original 5-pyramoids. (When the base *is* a regular $2n$-gon, of course, one *can* use a 5-pyramoid congruent to the original one, creating a polyhedron with S_{2nv} symmetry, but most of the time this will not be the case.) So one will get gyrobipyramoids like the one illustrated in Figure 45 below, where one can easily see that the triangular faces on the upper 5-pyramoid are much more acute than on the lower, and the pentagonal faces are quite different in appearance as well, with the upper pentagons having a near-right angle (actually approximately 87.978°) at the base, while the lower pentagons have significantly more obtuse angles.

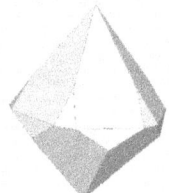

Figure 45: A hexagonal gyrobipyramoid.

Another possible polyhedron that can be constructed starts with the 5-pyramoid and fuses the base to the $n\times2$-gonal base of a pyramid, or a frustum of a pyramid. (While pyramids, and their frusta, have mostly been considered whose bases are regular polygons, the bases need not be, and in this case, the fact that the 5-pyramoid has an $n\times2$-gonal base suggests the formation of a pyramid with an $n\times2$-gonal base to make the fusion possible. This produces a composite polyhedron that is not very interesting, but is mentioned here for completeness.

A more interesting composite polyhedron, because, by aligning faces in one plane and combining them, a simple polyhedron can be obtained from it, starts by noting, as was mentioned earlier, that the base of a 5-pyramoid is a $n \times 2$-gon, rather than a regular $2n$-gon. If n is odd (and only if n is odd), one can conceive of two such $n \times 2$-gons oppositely directed in parallel planes, so that the vertices of one are related to those of the other by inversion through the point halfway between the centers of the two polygons (see p. 7, Chapter 1 for the definition of *inversion through a point*). Orienting two congruent 5-pyramoids so their bases are on these two $n \times 2$-gons, and connecting the nearest vertices of the bases, produces a polyhedron which has not been met with in the other books on polyhedra which I have seen, but which has some real beauty, because it comes closer, for any given n, to a sphere than almost any other figure of S_{2nv} symmetry, and so I call it a *globoid*. (While this process was discussed only for odd n, a globoid can be constructed with even n as well; the only thing is that the vertices of this globoid, other than the apex and antiapex, will not be related by inversion through a point.) It could also, technically, be called a *gyroelongated bi-5-pyramoid*, but that is an awkward name. The two $n \times 2$-gons which were the bases of the 5-pyramoids and the edges that connect them form a quasiprism, as illustrated in Figure 46, and so the globoid is really a composite polyhedron consisting of a fusion of a quasiprism and two 5-pyramoids, which is illustrated in Figure 47. (Note that, while we were using hexagonal 5-pyramoids and their fusions as the basis for the previous figures, the fact that this construction only works for odd n means we will not be able to use them here. So Figure 47 will be constructed with heptagonal 5-pyramoids.)

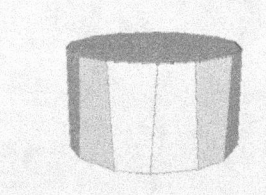

Figure 46: A bicropped heptagonal antiprism.

Figure 47: A heptagonal globoid.

The globoid is still a composite of three polyhedra, but it can be seen that if the height of the 5-pyramoids, the location of the apices of the triangular faces of the 5-pyramoids, and the height of the quasiprism are properly adjusted, the dihedral angles between the triangular faces and the bases of the 5-pyramoids can be adjusted, as can the dihedral angle between the isosceles trapezoidal faces and the bases of the quasiprism, so they sum to 180° exactly, and each triangle can be merged with an isosceles trapezoid (at the larger base of the trapezoid) into a single isosceles

Chapter 23. Apical polyhedra with pentagons at the apex I. General considerations.

pentagonal face. When this is done, *all* the faces of this polyhedron become pentagons, all isosceles, but not necessary congruent. Through all the operations of fusing a quasiprism with two 5-pyramoids, adjusting the dihedral angles to sum to 180°, and erasing the edges separating each triangle from the adjacent isosceles trapezoid, the S_{2nv} symmetry has been preserved, so this figure, which has $4n$ pentagonal faces, retains the S_{2nv} symmetry. Since it is derived from a globoid by eliminating the triangle/isosceles trapezoid edges, and has isosceles pentagons for all its faces, I call it a *pentagonized globoid*. There are $2n$ pentagonal faces that form two sets of n, one set meeting at the apex and one set at the antiapex, all congruent and all isosceles. And there are $2n$ pentagonal faces that form a lateral belt between the two, also all congruent and all isosceles, but not necessarily congruent to the pentagons of the first set. And here is the surprise. While "pentagonized globoid" is a new term for a whole family of polyhedra with S_{2nv} symmetry, *one* member of this family is very familiar. If n is made 3, *all twelve* of these pentagons can be made regular, and the symmetry becomes I_h. Yes, the triangular pentagonized globoid, if the angles are properly chosen, is the Platonic solid, the regular dodecahedron! An example of a heptagonal pentagonized globoid is shown below in Figure 48.

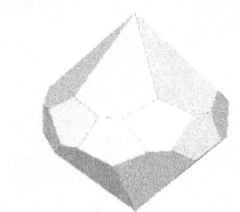

Figure 48: A heptagonal pentagonized globoid.

It should be noted that in constructing Figure 48, no attempt was made to produce any figure resembling the one in Figure 47, although it would be possible, by simply adjusting the relationship between the height of the quasiprism and the total height of the globoid, to make it a pentagonized globoid.

A summary of the polyhedra introduced in this chapter is given below in Table 12. The order of the principal axis of rotation is, as usual, denoted by n.

Name	Vertices					Edges	Faces				
	Order 3	Order 4	Order 5	Order n	Total		Triangles	Quadrilaterals	Pentagons	Others	Total
5-pyramoid	$3n$	0	0	1	$3n+1$	$5n$	n	0	n	1 $2n$-gon	$2n+1$
Orthobipyramoid	$2n$	$2n$	0	2	$4n+2$	$8n$	$2n$	0	$2n$	0	$4n$
Gyrobipyramoid	$2n$	$2n$	0	2	$4n+2$	$8n$	$2n$	0	$2n$	0	$4n$
Peritruncated bipyramid	$4n$	0	0	2	$4n+2$	$7n$	0	n (r)	$2n$	0	$3n$
Globoid	$2n$	$4n$	0	2	$6n+2$	$12n$	$2n$	$2n$ (IT)	$2n$	0	$6n$
Pentagonized globoid	$6n$	0	0	2	$6n+2$	$10n$	0	0	$4n$	0	$4n$

Table 12: Apical polyhedra with pentagons at the apex.

Note that in Table 12, (r) after the number of quadrilaterals means that they are all rhombi; and (IT) means they are all isosceles trapezoids. Of course, when $n = 3$ or $n = 4$, the "Order 3" or "Order 4" column must be combined with the "Order n" column, so that a triangular peritruncated bipyramid, for example has 14 (= $4n + 2$) threefold vertices.

Chapter 24. Apical polyhedra with pentagons at the apex II. Mathematical details. *[Optional]*

The first thing that will be computed in this chapter will be the coordinates of the vertices of a 5-pyramoid. Because the other polyhedra described in the previous chapter are all derived from the 5-pyramoid by fusion, this will serve as a first step toward the determination of the vertices of each of the other types of apical polyhedron with pentagons at the apex. All of the appropriate conventions that have been followed as standard in this book will be retained:

The order of the principal rotation axis of the polyhedron will be designated as n, and the angle α is equal to $360°/n$, or $2\pi/n$ in radians.

Vertex rotation plane based labeling is used to designate the vertices: the apex A will be assigned the coordinates $(0, 0, z_A)$, where z_A will be, as yet, arbitrary. The points B_k, where $k = 1, 2, ..., n$, in the second vertex rotation plane, are given the coordinates $(r_B \cos \{k-1\}\alpha, r_B \sin \{k-1\}\alpha, z_B)$, where α is equal to $360°/n$, or $2\pi/n$ in radians. Again, both r_B and z_B can be considered arbitrary, although according to our convention for the z-coordinate, where the apex is the highest-z vertex, $z_B < z_A$. The points C_j, where $j = 1, 2, ..., 2n$, in the third vertex rotation plane, fall into two groups, depending on whether j is odd or even. If j is *odd*, one can write $j = 2k - 1$, and the points $C_j = C_{2k-1}$ are given the coordinates $(r_C \cos \{[k-1]\alpha + \beta\}, r_C \sin \{[k-1]\alpha + \beta\}, z_C)$. If j is *even*, one can write $j = 2k$, and the points $C_j = C_{2k}$ are given the coordinates $(r_C \cos \{k\alpha - \beta\}, r_C \sin \{k\alpha - \beta\}, z_C)$. The conditions need to be determined that will make the five vertices of each of the pentagonal faces $AB_kC_{2k-1}C_{2k}B_{k+1}$ coplanar. The vector methods of Appendix B will be convenient for this purpose. It is only necessary to look at one of the pentagons, namely $AB_1C_1C_2B_2$, since all the others are related by symmetry. It will not be necessary to do anything to ensure that triangles such as $B_2C_2C_3$ are planar, because *all* sets of three points are coplanar. We have determined the coordinates of the 5-pyramoid as follows:

$A: (0, 0, z_A)$.

$B_1: (r_B, 0, z_B)$.

$B_2: (r_B \cos \alpha, r_B \sin \alpha, z_B)$.

$C_1: (r_C \cos \beta, r_C \sin \beta, z_C)$.

$C_2: (r_C \cos [\alpha - \beta], r_C \sin [\alpha - \beta], z_C)$.

To simplify the form of the algebraic expressions, let $s_1 = \sin \alpha$, $s_2 = \sin \beta$, $c_1 = \cos \alpha$, $c_2 = \cos \beta$, $h_1 = z_A - z_B$, $h_2 = z_A - z_C$. Then:

$\cos (\alpha - \beta) = c_1 c_2 + s_1 s_2$ and

$\sin (\alpha - \beta) = s_1 c_2 - s_2 c_1$,

which means that the coordinates of B_2, C_1, and C_2 can be written as:

$B_2: (r_B c_1, r_B s_1, z_B)$.

$$C_1: (r_C c_2, r_C s_2, z_C).$$
$$C_2: (r_C(c_1 c_2 + s_1 s_2), r_C(s_1 c_2 - s_2 c_1), z_C).$$

Using the formulas of Appendix B, the coordinates of the midpoint of $B_1 B_2$ can be computed as:
$$M_{B12}: (r_B (1+c_1)/2, r_B s_1/2, z_B),$$
which means that the vector from A to that midpoint is:
$$AM_{B12} = (r_B (1+c_1)/2, r_B s_1/2, -h_1).$$

Similarly, the coordinates of the midpoint of $C_1 C_2$ can be computed as
$$M_{C12}: (r_C[c_2 + (c_1 c_2 + s_1 s_2)]/2, r_C[s_2 + (s_1 c_2 - s_2 c_1)]/2, z_C),$$
and thus the vector from A to that midpoint is:
$$AM_{C12} = (r_C[c_2 + (c_1 c_2 + s_1 s_2)]/2, r_C[s_2 + (s_1 c_2 - s_2 c_1)]/2, -h_2).$$

The simplest way to assure that the points A, B_1, B_2, C_1, and C_2 are coplanar, so that a planar pentagon $AB_1 C_1 C_2 B_2$ is obtained, is is to assure that these two midpoints M_{B12} and M_{C12} are on a line with A, because it can easily be shown that $B_1 B_2$ and $C_1 C_2$ are parallel. This in turn requires the vectors AM_{B12} and AM_{C12} to be related by a proportionality:
$$AM_{B12}/h_1 = AM_{C12}/h_2,$$
or expanding in components,
$$h_2 r_B(1+c_1)/2 = h_1 r_C[c_2 + (c_1 c_2 + s_1 s_2)]/2,$$
$$h_2 r_B s_1/2 = h_1 r_C[s_2 + (s_1 c_2 - s_2 c_1)]/2.$$

Simplifying these equations, and temporarily letting $K = h_2 r_B / h_1 r_C$,
$$K(1 + c_1) = c_2 + c_1 c_2 + s_1 s_2,$$
$$K s_1 = s_2 + s_1 c_2 - s_2 c_1.$$

Since the same value of K must apply to both equations, we must have:
$$s_1(c_2 + c_1 c_2 + s_1 s_2) = (1 + c_1)(s_2 + s_1 c_2 - s_2 c_1), \text{ or}$$
$$s_1 c_2 + s_1 c_1 c_2 + s_1^2 s_2 = s_2 + s_1 c_2 + s_1 c_1 c_2 - s_2 c_1^2;$$
$$s_1^2 s_2 = s_2 - s_2 c_1^2;$$
$$s_1^2 = 1 - c_1^2.$$

But this is true whatever the value of α is, since this is a standard identity involving sines and cosines. So one needs to go back to either one of the two equations that were solved together to give this result. We will choose the simpler of the two:
$$h_2 r_B s_1/2 = h_1 r_C[s_2 + (s_1 c_2 - s_2 c_1)]/2,$$
which can be solved to give a relationship between h_1, h_2, r_B, and r_C:
$$h_2 r_B s_1 = h_1 r_C[s_2 + (s_1 c_2 - s_2 c_1)],$$
$$h_2 r_B = h_1 r_C[s_2 + (s_1 c_2 - s_2 c_1)]/s_1.$$

Chapter 24. Apical polyhedra with pentagons at the apex II. Mathematical details. [Optional]

$$h_2 r_B / h_1 r_C = [s_2 + (s_1 c_2 - s_2 c_1)]/s_1,$$

or, putting back in the quantities that were originally used in this computation,

$$(z_A - z_C) r_B / (z_A - z_B) r_C = [\sin \beta + \sin(\alpha - \beta)]/\sin \alpha.$$

This equation gives a lot of freedom. There are seven quantities appearing in this equation, namely r_B, r_C, z_A, z_B, z_C, α, and β, and from any six of these, the seventh can be computed. There are some requirements that must be met, such as that $\beta < \alpha/2$ and that the values of the quantities on the left-hand side do not cause one of the sines on the right-hand side to exceed 1. But this still leaves an extraordinary degree of freedom.

This polyhedron can also be constructed by considering it in a different way. If one starts with a pyramid and truncates all the vertices except the apex, doing it in a symmetric way so that the pieces which are cut off at each vertex of the base are congruent and symmetrically arranged about the mirror plane that passes through the vertex in question, the figure obtained is a 5-pyramoid of the same nature as the one that has been described. And because the pentagonal faces are obtained by truncating triangular faces of the starting pyramid, the vertices of those faces will automatically be coplanar. For this purpose, the vertices A, B_k, and C_k will be defined before, but a set of original vertices of the pyramid need to be fixed. They will be given coordinates

$$P_k: (r_P \cos\{k-1\}\alpha, \; r_P \sin\{k-1\}\alpha, \; z_C),$$

where it is already noted that the z-coordinate will be the same as that of all the C_k. It will be necessary to locate the points B_k and C_k, where B_k is along the line AP_k and each C_j, if j is odd and $j=2k-1$, is located along the line $P_k P_{k+1}$ but closer to P_k, while if j is even and $j=2k$, it is located along the line $P_k P_{k+1}$ but closer to P_{k+1}. Furthermore, all the distances AB_k are equal, and all the distances $P_k C_{2k-1}$ and $C_{2k} P_{k+1}$ are also equal (and less than the distances $P_k P_{k+1}$).

Because of symmetry, again, one only needs to look at the points A, B_1, B_2, C_1, and C_2, which can be calculated from the coordinates of A, P_1, and P_2, where we have, simplifying the above formulas:

$$P_1: (r_P, 0, z_C),$$
$$P_2: (r_P \cos \alpha, \; r_P \sin \alpha, \; z_C).$$

Again, the formulas of Appendix B are useful. They determine the coordinates of points C_1 and C_2 as follows:

$$C_1: ([1 - f + f \cos \alpha] r_P, \; f r_P \sin \alpha, \; z_C)$$
$$= ([1 - f[1 - \cos \alpha]] r_P, \; f r_P \sin \alpha, \; z_C),$$
$$C_2: (\{f + [1 - f] \cos \alpha\} r_P, \; [1 - f] r_P \sin \alpha, \; z_C)$$
$$= (\{\cos \alpha + f[1 - \cos \alpha]\} r_P, \; [1 - f] r_P \sin \alpha, \; z_C),$$

where f is the fraction of the way from one point to the other, i.e.,

$$P_1C_1 = fP_1P_2,$$
$$C_2P_2 = fP_1P_2.$$

All the other points C_k can be obtained by rotating C_1 and C_2 around the z-axis through multiples of α.

To simplify the algebra, let

$$x_1 = [1 - f(1 - \cos\alpha)]r_P,$$
$$y_1 = fr_P \sin\alpha,$$
$$x_2 = [\cos\alpha + f(1 - \cos\alpha)]r_P, \text{ and}$$
$$y_2 = (1 - f)r_P \sin\alpha.$$

Then we can write

$$C_1: (x_1, y_1, z_C),$$
$$C_2: (x_2, y_2, z_C),$$

giving

$$C_{2k-1}: (x_1 \cos[k-1]\alpha - y_1 \sin[k-1]\alpha, x_1 \sin[k-1]\alpha + y_1 \cos[k-1]\alpha, z_C),$$
$$C_{2k}: (x_2 \cos[k-1]\alpha - y_2 \sin[k-1]\alpha, x_2 \sin[k-1]\alpha + y_2 \cos[k-1]\alpha, z_C),$$

with $k = 1, 2, ..., n$. It is not obvious from this that the two points C_{2k-1} and C_{2k} are equidistant from the z-axis, as is automatic from the first construction, though from the symmetry of the problem, they ought to be. But calculating

$$x_1^2 + y_1^2 = \{[1 - f(1 - \cos\alpha)]r_P\}^2 + (fr_P \sin\alpha)^2$$
$$= \{[1 - f(1 - \cos\alpha)]^2 + (f \sin\alpha)^2\}r_P^2$$
$$= [(1 - f + f\cos\alpha)^2 + (f\sin\alpha)^2]r_P^2$$
$$= [(1 + f^2 + f^2\cos^2\alpha - 2f + 2f\cos\alpha - 2f^2\cos\alpha) + (f^2\sin^2\alpha)]r_P^2$$
$$= (1 - 2f + 2f\cos\alpha + f^2 - 2f^2\cos\alpha + f^2\cos^2\alpha + f^2\sin^2\alpha)r_P^2$$
$$= (1 - 2f + 2f\cos\alpha + f^2 - 2f^2\cos\alpha + f^2)r_P^2$$
$$= (1 - 2f + 2f\cos\alpha + 2f^2 - 2f^2\cos\alpha)r_P^2$$
$$= [1 - 2f(1 - \cos\alpha) + 2f^2(1 - \cos\alpha)]r_P^2$$
$$= [1 - 2f(1 - f)(1 - \cos\alpha)]r_P^2$$

$$x_2^2 + y_2^2 = \{[\cos\alpha + f(1 - \cos\alpha)]r_P\}^2 + [(1 - f)r_P \sin\alpha]^2$$
$$= \{[\cos\alpha + f(1 - \cos\alpha)]^2 + [(1 - f)\sin\alpha]^2\}r_P^2$$
$$= [(\cos\alpha + f - f\cos\alpha)^2 + (1 - 2f + f^2)\sin^2\alpha]r_P^2$$
$$= [(\cos^2\alpha + f^2 + f^2\cos^2\alpha + 2f\cos\alpha - 2f\cos^2\alpha - 2f^2\cos\alpha) + (\sin^2\alpha - 2f\sin^2\alpha + f^2\sin^2\alpha)]r_P^2$$
$$= (\cos^2\alpha + \sin^2\alpha + 2f\cos\alpha - 2f\cos^2\alpha - 2f\sin^2\alpha + f^2 + f^2\sin^2\alpha + f^2\cos^2\alpha - 2f^2\cos\alpha)r_P^2.$$

Chapter 24. Apical polyhedra with pentagons at the apex II. Mathematical details. [Optional]

Replacing $\cos^2 \alpha + \sin^2 \alpha$ by 1, we have:

$$x_2^2 + y_2^2 = (1 + 2f\cos\alpha - 2f + 2f^2 - 2f^2\cos\alpha)r_P^2$$
$$= [1 - 2f(1 - \cos\alpha) + 2f^2(1 - \cos\alpha)]r_P^2$$
$$= [1 - 2f(1 - f)(1 - \cos\alpha)]r_P^2.$$

This is, of course, the same result obtained for $x_1^2 + y_1^2$, as it should be, confirming the requirement of symmetry.

The point B_1 can be located anywhere along the line AP_1, so denoting the fraction AB_1/AP_1 by q, one has the coordinates of B_1 given as:

$$B_1: (qr_P, 0, \{[1 - q]z_A + qz_C\})$$

If we want to use z_B for the z-coordinate of the B_k points, we must have:

$$z_B = (1 - q)z_A + qz_C$$
$$= z_A - q(z_A - z_C);$$
$$q(z_A - z_C) = z_A - z_B;$$
$$q = (z_A - z_B)/(z_A - z_C).$$

The two recipes for constructing the 5-pyramoid should be compared. In the first, we had the coordinates

$$A: (0, 0, z_A),$$
$$B_k: (r_B \cos \{k - 1\}\alpha, r_B \sin \{k - 1\}\alpha, z_B) \text{ where } k = 1, 2, ..., n,$$
$$C_{2k-1}: (r_C \cos \{[k - 1]\alpha + \beta\}, r_C \sin \{[k - 1]\alpha + \beta\}, z_C),$$
$$C_{2k}: (r_C \cos \{k\alpha - \beta\}, r_C \sin \{k\alpha - \beta\}, z_C),$$

where

$$\alpha = 360°/n, \text{ or } 2\pi/n \text{ in radians,}$$
$$z_C < z_B < z_A, \text{ and}$$
$$(z_A - z_C)r_B/(z_A - z_B)r_C = [\sin\beta + \sin(\alpha - \beta)]/\sin\alpha.$$

From any six of the seven quantities r_B, r_C, z_A, z_B, z_C, n, and β, the seventh can be computed (though n will normally be one of the six quantities taken, rather than computed, since it must be an integer). In the second, we had the same choice for the apex A, but we found

$$B_k: (qr_P \cos \{k - 1\}\alpha, qr_P \sin \{k - 1\}\alpha, z_B),$$
$$C_1: (\{1 - f[1 - \cos\alpha]\}r_P, fr_P \sin\alpha, z_C),$$
$$C_2: (\{\cos\alpha + f[1 - \cos\alpha]\}r_P, [1 - f]r_P \sin\alpha, z_C),$$

and the other C_k determined by rotating C_1 and C_2 through multiples of α, where

$$\alpha = 360°/n, \text{ or } 2\pi/n \text{ in radians,}$$
$$0 < f < 1,$$
$$z_C < z_B < z_A, \text{ and}$$
$$q = (z_A - z_B)/(z_A - z_C).$$

Here again, one can choose z_A, z_B, z_C, and n, but instead of picking two of the three quantities r_B, r_C, and β, we have the two quantities f and r_P to choose. Of course they can be related, because both recipes give the same 5-pyramoid coordinates. If we started with f and r_P and wish to determine r_B, r_C, and β, we can use the result above for $x_1^2 + y_1^2$ (or $x_2^2 + y_2^2$) to determine r_C by $r_C^2 = x_1^2 + y_1^2 = [1 - 2f(1-f)(1 - \cos\alpha)]r_P^2$. Then we have $r_B = qr_P = (z_A - z_B)r_P/(z_A - z_C)$, and finally β can be computed from the equation
$$(z_A - z_C)r_B/(z_A - z_B)r_C = [\sin\beta + \sin(\alpha - \beta)]/\sin\alpha,$$
just as if we had chosen r_B and r_C as the two of the three quantities in the first recipe to start. Of course, if we had started with two of the three quantities r_B, r_C, and β, we could similarly compute f and r_P, since we have $r_B = qr_P$, which can be rearranged to give
$$r_P = r_B/q$$
$$= (z_A - z_C)r_B/(z_A - z_B),$$
though getting f from the equation
$$r_C^2 = [1 - 2f(1-f)(1 - \cos\alpha)]r_P^2$$
is not a simple task, involving solving a quadratic equation. It is certainly possible, though.

Since both recipes give the same results except for the choice of quantities to start with, further discussion in this chapter will assume the use of whichever recipe is convenient.

It is relatively easy to calculate the dihedral angle between the triangular faces and the base; it will be the same as the angle B_1MO_C, where M is the midpoint of the edge C_1C_{2n} and O_C is where the z-axis cuts the vertex rotation plane including all the points C_k. If we use the first recipe, the coordinates of B_1 are given by $(r_B, 0, z_B)$, those of O_C by $(0, 0, z_C)$, and since those of C_1 are given by $(r_C \cos\beta, r_C \sin\beta, z_C)$ and C_{2n} by $(r_C \cos\beta, -r_C \sin\beta, z_C)$, the coordinates of M are given by $(r_C \cos\beta, 0, z_C)$. Then:

$$MB_1 = (r_B - r_C \cos\beta, 0, z_B - z_C)$$
$$|MB_1| = \sqrt{[(r_B - r_C \cos\beta)^2 + (z_B - z_C)^2]}$$
$$= \sqrt{[r_B^2 - 2r_Br_C \cos\beta + r_C^2 \cos^2\beta + (z_B - z_C)^2]}$$
$$MO_C = (-r_C \cos\beta, 0, 0)$$
$$|MO_C| = r_C \cos\beta$$
$$MB_1 \cdot MO_C = r_C^2 \cos^2\beta - r_Br_C$$
$$\cos\delta = MB_1 \cdot MO_C/(|MB_1||MO_C|),$$

with the expressions given for $|MB_1|$, $|MO_C|$, and $MB_1 \cdot MO_C$. However, most of the time only the *sign* of the cosine of the dihedral angle δ is necessary, and that will be the

Chapter 24. Apical polyhedra with pentagons at the apex II. Mathematical details. [Optional]

same as the sign of $\mathbf{MB_1} \cdot \mathbf{MO_G}$, which in turn will be the same as the sign of $r_C \cos^2 \beta - r_B$, since r_C is always positive. If $r_C \cos^2 \beta > r_B$, δ will be acute. If $r_C \cos^2 \beta = r_B$, δ will be right And if $r_C \cos^2 \beta < r_B$, δ will be obtuse.

To form an *orthobipyramoid* or a *peritruncated bipyramid*, one simply starts as one did for the 5-pyramoid and computes a set of coordinates of points A, B_k (k=1 to n), and C_k (k=1 to $2n$) by either of the preceding recipes. The additional vertices D_k (k=1 to n) and E are computed rather simply: whatever the x- and y-coordinates of B_k are, the same x- and y-coordinates are used for D_k; but the common z-coordinate for all the D_k is given by $z_D = 2z_C - z_B$; in the same way one fixes E: $(0, 0, z_E)$, with $z_E = 2z_C - z_A$. This will be a symmetrical orthobipyramoid if the dihedral angle at the base is acute, implying $r_C \cos^2 \beta > r_B$, and a peritruncated bipyramid if it is right, which means that $r_C \cos^2 \beta = r_B$; if it is obtuse, of course, the combination will not produce a convex polyhedron, so that choice of starting parameters cannot be used.

Forming a gyrobipyramoid, however, is more difficult. Because the base of a 5-pyramoid is not, in general, a regular $2n$-gon but rather an $n \times 2n$-gon, one cannot, normally, fit together two congruent 5-pyramoids with the triangular faces of one sharing edges with the pentagonal faces of the other. There are two possible ways to get around the problem.

First, if the parameters are properly chosen (specifically, if β = α/4) the base polygon *will* be a regular $2n$-gon. In that case the gyrobipyramoid will have S_{2nv} symmetry and the coordinates can be derived very simply by the use of this symmetry. One can compute $z_D = 2z_C - z_B$ and $z_E = 2z_C - z_A$, fix E at $(0, 0, z_E)$, and the various D_k points by rotating the x and y coordinates of the corresponding B_k points by α/2. The details of this will not be given here, but can easily be produced by anyone who has come this far in the chapter.

Second, the bottom 5-pyramoid can be *generated to fit the base*. For this purpose, the second recipe will be most convenient. As it started with a set of n points, all in one plane, which formed the basal vertices of a pyramid that was truncated to form the 5-pyramoid, a different pyramid, inverted in orientation, can be built to start. Recall that C_1C_2 was a fragment of the original P_1P_2; C_3C_4, of the original P_2P_3, and so forth. To get the new pyramid oriented properly, its base edges must lie along C_2C_3, C_4C_5, ..., $C_{2n}C_1$. Since these are all in one plane, it is relatively easy to compute the vertices of this new pyramid: P_1', for example, is the intersection of C_2C_3 with C_4C_5. These new P_k' can all be computed simply; again the details of this will not be given here, but the reader can complete the task if he has come this far in the chapter.

To form a *globoid*, however, the inequality of the basal edges causes no problem. The quasiprism that is to be fused with the 5-pyramoid can be chosen to have its edges alternate the same way, and the fusion is perfect; the longer edges of the top of the quasiprism align like the shorter edges of the bottom, and vice versa, so two identical 5-pyramoids can be fused to the quasiprism, rotated by $\alpha/2$ and inverted, relative to each other. It might be noted that if the 5-pyramoids are chosen so that each base is a regular $2n$-gon, a prism, rather than a quasiprism, is needed to fit the two; the resulting polyhedron is perhaps not exactly in accordance with the definition of a globoid in the previous chapter, but probably ought to be considered a globoid as well, possibly with a special qualification such as a *regular* globoid.

We will use a slightly different set of coordinates in constructing a globoid, however, in order to take advantage of the S_{2nv} symmetry. We will locate the origin exactly halfway between the apex and the antiapex, so that the z-coordinates of A (the apex) and F (the antiapex) are equal except for the sign. And similarly, with this choice of coordinates, $z_B = -z_E$ and $z_C = -z_D$. Giving α and β the same significance as in our first recipe, we can therefore determine the coordinates as follows:

$$A: (0, 0, z_A)$$
$$B_k: (r_B \cos [k-1]\alpha, r_B \sin [k-1]\alpha, z_B)$$
$$C_{2k-1}: (r_C \cos \{[k-1]\alpha + \beta\}, r_C \sin \{[k-1]\alpha + \beta\}, z_C)$$
$$C_{2k}: (r_C \cos [k\alpha - \beta], r_C \sin [k\alpha - \beta], z_C)$$
$$D_{2k-1}: (r_C \cos \{[k-\tfrac{1}{2}]\alpha - \beta\}, r_C \sin \{[k-\tfrac{1}{2}]\alpha - \beta\}, -z_C)$$
$$D_{2k}: (r_C \cos \{[k-\tfrac{1}{2}]\alpha + \beta\}, r_C \sin \{[k-\tfrac{1}{2}]\alpha + \beta\}, -z_C)$$
$$E_k: (r_B \cos [k-\tfrac{1}{2}]\alpha, r_B \sin [k-\tfrac{1}{2}]\alpha, -z_B)$$
$$F: (0, 0, -z_A)$$

Here, still, we require $z_C < z_B < z_A$ and $\beta < \tfrac{1}{2}\alpha$, as before. Because of the symmetry, it is necessary only to work with one pentagon, such as $AB_1C_1C_2B_2$, one triangle, such as $B_1C_1C_{2n}$, and one isosceles trapezoid, such as $C_1C_{2n}D_{2n}D_1$. All the others will be congruent to these by symmetry. Nothing need really be done with triangle $B_1C_1C_{2n}$, except that we might wish to determine some angles and edge lengths:

$$B_1: (r_B, 0, z_B),$$
$$C_1: (r_C \cos \beta, r_C \sin \beta, z_C), \text{ and}$$
$$C_{2n}: (r_C \cos \beta, -r_C \sin \beta, z_C),$$

noting that $\cos 0° = 1$, $\sin 0° = 0$, $\cos (n\alpha - \beta) = \cos (360° - \beta) = \cos \beta$, and $\sin (n\alpha - \beta) = \sin (360° - \beta) = -\sin \beta$, from the definition of α, simplifying the formulas, giving the edge lengths

$$B_1C_1 = B_1C_{2n} = \sqrt{[(r_B - r_C \cos \beta)^2 + r_C^2 \sin^2 \beta + (z_B - z_C)^2]} \text{ and}$$
$$C_1C_{2n} = 2r_C \sin \beta.$$

The angle $C_{2n}B_1C_1$ can be determined by the vector methods of Appendix B:

$$B_1C_1 \cdot B_1C_{2n} = (r_B - r_C \cos \beta)^2 - r_C^2 \sin^2 \beta + (z_B - z_C)^2$$

Chapter 24. Apical polyhedra with pentagons at the apex II. Mathematical details. [Optional]

$$\cos C_{2n}B_1C_1 = \mathbf{B_1C_1} \cdot \mathbf{B_1C_{2n}}/|\mathbf{B_1C_1}||\mathbf{B_1C_{2n}}| =$$
$$[(r_B - r_C \cos \beta)^2 - r_C^2 \sin^2 \beta + (z_B - z_C)^2]/[(r_B - r_C \cos \beta)^2 + r_C^2 \sin^2 \beta + (z_B - z_C)^2].$$

This is always less than 1 as long as $(r_B - r_C \cos \beta)^2 + (z_B - z_C)^2 > r_C^2 \sin^2 \beta$, since the term $r_C^2 \sin^2 \beta$ is always positive, so the value on the right hand side is a legitimate value for the cosine of an angle.

The four vertices C_1, C_{2n}, D_1, and D_{2n} will always be coplanar. This can be verified by showing that the two vectors $\mathbf{C_1C_{2n}}$ and $\mathbf{D_1D_{2n}}$ are parallel, making $C_1C_{2n}D_{2n}D_1$ a trapezoid. In fact they are both parallel to the y-axis:

C_1: $(r_C \cos \beta, r_C \sin \beta, z_C)$,
C_{2n}: $(r_C \cos \beta, -r_C \sin \beta, z_C)$,
$\mathbf{C_1C_{2n}} = (0, -2r_C \sin \beta, 0)$; and
D_1: $(r_C \cos [\alpha/2 - \beta], r_C \sin [\alpha/2 - \beta], -z_C)$,
D_{2n}: $(r_C \cos [\alpha/2 - \beta], -r_C \sin [\alpha/2 - \beta], -z_C)$,
$\mathbf{D_1D_{2n}} = (0, -2r_C \sin [\alpha/2 - \beta], 0)$.

So there is no need to do anything to ensure planarity of the isosceles trapezoid $C_1C_{2n}D_{2n}D_1$. That it *is* isosceles follows from the symmetry of the coordinates of C_1 and C_{2n}, as well as of the coordinates of D_1 and D_{2n}.

It is only necessary, then, to make certain that the vertices of pentagon $AB_1C_1C_2B_2$ lie in a plane.

A: $(0, 0, z_A)$,
B_1: $(r_B, 0, z_B)$,
B_2: $(r_B \cos \alpha, r_B \sin \alpha, z_B)$,
C_1: $(r_C \cos \beta, r_C \sin \beta, z_C)$, and
C_2: $(r_C \cos [\alpha - \beta], r_C \sin [\alpha - \beta], z_C)$.

But this was exactly the same problem we had in the first recipe for the 5-pyramoid, so the solution is exactly the same: $AB_1C_1C_2B_2$ will be planar if

$$(z_A - z_C)r_B/(z_A - z_B)r_C = [\sin \beta + \sin (\alpha - \beta)]/\sin \alpha.$$

So the globoid requires exactly the same conditions as the original 5-pyramoid, which can be interpreted as meaning that from any 5-pyramoid, a globoid can be constructed by the fusion of it and another globoid, congruent to it, to the bases of a bicropped antiprism. (Actually this will not always be true, because to ensure convexity the dihedral angles at the C vertex rotation plane [and, by symmetry, the D vertex rotation plane] must all be less than 180°.)

To make a *pentagonized globoid*, one additional condition must be satisfied, but otherwise the same construction as for the globoid just described suffices. It is necessary for the vertices B, C_1, D_1, D_{2n}, and C_{2n} to be coplanar, forming a pentagon. Because of symmetry, all the other required pentagons will automatically be obtained. We follow a procedure similar to the one used to ensure the coplanarity of the vertices forming the pentagonal faces of the 5-pyramoid, namely, determine the midpoints of $C_1 C_{2n}$ and $D_1 D_{2n}$, and make certain that these two midpoints and B are all on a single line. We already determined the coordinates:

$$B_1: (r_B, 0, z_B),$$
$$C_1: (r_C \cos \beta, r_C \sin \beta, z_C),$$
$$C_{2n}: (r_C \cos \beta, -r_C \sin \beta, z_C),$$
$$D_1: (r_C \cos [\alpha/2 - \beta], r_C \sin [\alpha/2 - \beta], -z_C), \text{ and}$$
$$D_{2n}: (r_C \cos [\alpha/2 - \beta], -r_C \sin [\alpha/2 - \beta], -z_C),$$

so all we need to calculate are the coordinates of the midpoints $M_{C_1,2n}$ and $M_{D_1,2n}$. This is simple, given the symmetry:

$$M_{C_1,2n}: (r_C \cos \beta, 0, z_C),$$
$$M_{D_1,2n}: (r_C \cos [\alpha/2 - \beta], 0, -z_C),$$

and because all the y-coordinates of B, $M_{C_1,2n}$, and $M_{D_1,2n}$ are 0, the condition for collinearity is simply

$$(r_B - r_C \cos \beta)/(z_B - z_C) = [\cos \beta - \cos (\alpha/2 - \beta)]r_C/2z_C.$$

Since this is another equation in r_B, r_C, z_B, z_C, α, and β, this reduces the degrees of freedom by one from that of the 5-pyramoid. Since α is determined by n, we can start with any five of the six quantities r_B, r_C, z_B, z_C, n, and β, and determine the sixth; then from the previous equation

$$(z_A - z_C)r_B/(z_A - z_B)r_C = [\sin \beta + \sin (\alpha - \beta)]/\sin \alpha,$$

determine z_A.

Chapter 25. A final discussion of apical polyhedra.

The last few chapters have focused on apical polyhedra, classified by the type of polygons which meet at the apex. All the Platonic solids were included as special types of the apical polyhedra discussed there, and this classification deserves a summary table.

Name	Type of polyhedron	Class of polyhedron family	Order of principal axis of rotation
Tetrahedron	Pyramid	Apicobasal	3
Octahedron	Bipyramid	Diapical	4
Icosahedron	Biaugmented anti-prism	Diapical	5
Hexahedron or cube	Antibipyramid	Diapical	3
Dodecahedron	Pentagonized globoid	Diapical	3

Table 13: The Platonic solids, considered as apical polyhedra.

We did not go beyond pentagons as the chosen polygons meeting at the apex; this was only because, by going this far, we obtained the generalizations of all the Platonic solids. But this is not the end of the progression. One could, for example, construct a 6-pyramoid such as Figure 49 below.

Figure 49: A square 6-pyramoid.

We have also restricted ourselves to apical polyhedra with only *one* type of polygon at the apex. This is not a necessary limitation; in fact, one could start with any of the Archimedean solids in which an axis of rotation can be found through a vertex, such as the cuboctahedron with its twofold axis through any pair of opposite vertices, and generalize it. Each vertex has an equilateral triangle alternating with a square; uniaxial stretching along an axis through two opposite vertices produces a diapical polyhedron in which the apex and antiapex have a twofold alternation of isosceles triangles and rhombi, while the faces that were, in the original cuboctahedron, squares not having a vertex at the two polyhedron vertices selected for the apex and antiapex become rectangles (see Figure 50 below), and the same sort of generalization that enables us to consider, say, the octahedron as a special case of a bipyramid can be used to build a polyhedron with a threefold, fourfold, etc., alternation of isosceles triangles and rhombi. But we do not have to stop at polyhedra which can be considered generalizations of Archimedean solids. As we constructed polyhedra such as orthobipyramoids, which are not generalizations of Platonic solids, we can use any combination of polygons (if we stretch them enough to produce an angle sum less than 180°) to produce new types of polyhedra. The only reason we do not, in this book, go any further is that we have a finite amount of time available, and the number of types of apical polyhedron with axial symmetry is infinite, so that we need to stop *somewhere*.

Figure 50: A cuboctahedron stretched along one of its twofold axes.

Actually, a better place than the Archimedean solids to start with, if one wishes to consider some polyhedron as the basis for a family by changing the order of a principal axis of rotation through a vertex, is the set of *duals* of the Archimedean solids, sometimes called the *Catalan solids*. The number of Archimedean solids with even a *twofold* axis through a vertex is small; by contrast, the Catalan solids *all* have axes of rotation passing through vertices (as can be deduced by noting that all the Archimedean solids, by definition, have regular polygons as faces, and applying the rules for relating duals to the original polyhedra). It can be seen that the Archimedean solids are more likely, by looking at principal axes of rotation through *face centers*, to produce families of *tectal* polyhedra with axial symmetry.

Yet another restriction we have observed in the preceding chapters on axial polyhedra is that of *maximal symmetry for the type*. If the polyhedron was apicobasal, the base was assumed to be a regular polygon (except in the case of the 5-pyramids, for which the sides that were shared edges with pentagonal faces were necessarily different from those that were shared edges with triangular faces, so it was unnecessary to make them equal, and the bases were $n \times 2$-gons rather than regular $2n$-gons.). This is

Chapter 25. A final discussion of apical polyhedra.

the least important of the restrictions. It was imposed to make it easier to compute the coordinates of the vertices of the polyhedra, but in the classification of the polyhedra it makes little difference. A pyramid whose base is a 2×3-gon or a 3×2-gon, after all, is still a hexagonal pyramid, just not a *regular* hexagonal pyramid.

But at this time it will be this author's plan to terminate the discussion of apical polyhedra with axial symmetry (though the reader is certainly not discouraged from studying some of the types we ignored on his own) and start discussing *tectal* polyhedra with axial symmetry. While we have implicitly treated some of the kinds in the past (prisms, prismoids, and antiprisms are among the members of that category of polyhedra, and the dual of a diapical polyhedron with axial symmetry is a tectal polyhedron with axial symmetry), it will be the intent of the next few chapters to treat tectal polyhedra with axial symmetry in about the same level of detail as we did the apical polyhedra.

Table 14 below, similar to the previous Tables 7 and 9, provides a summary of the symmetry properties of a number of apical polyhedra. (It might be noted that in some cases, the notation "S_{2nv} or C_{nv}" is found. These polyhedra are of symmetry group S_{2nv} when the pyram(o)id at the apex is identical with the one at the antiapex, and they are otherwise of symmetry group C_{nv}.)

Name of polyhedron	Symmetry group	n-Isohedral n	n-Isogonal n	n-Isotoxal n
pyramid	C_{nv}	2	2	2
4-pyramoid	C_{nv}	3	3	3
5-pyramoid	C_{nv}	3	3	4
6-pyramoid	C_{nv}	4	4	5
bipyramid	D_{nh}	1	2	1
antibipyramid	S_{2nv}	1	2	2
polykite	C_{nv}	2	4	3
polytimoid	C_{nv}	3	4	4
orthobipyramoid	D_{nh}	2	3	4
gyrobipyramoid	S_{2nv} or C_{nv}	2 (S_{2nv}) or 4 (C_{nv})	3 (S_{2nv}) or 6 (C_{nv})	3 (S_{2nv}) or 6 (C_{nv})
peritruncated bipyramid	D_{nh}	2	3	3
globoid	S_{2nv} or C_{nv}	3 (S_{2nv}) or 5 (C_{nv})	3 (S_{2nv}) or 6 (C_{nv})	5 (S_{2nv}) or 7 (C_{nv})
pentagonized globoid	S_{2nv}	2	3	3
monoaugmented antiprism	C_{nv}	4	3	4

Name of polyhedron	Symmetry group	n-Isohedral n	n-Isogonal n	n-Isotoxal n
pyramid	C_{nv}	2	2	2
biaugmented antiprism	S_{2nv} or C_{nv}	2 (S_{2nv}) or 4 (C_{nv})	2 (S_{2nv}) or 4 (C_{nv})	3 (S_{2nv}) or 5 (C_{nv})

Table 14: Primary symmetry properties of some apical polyhedra.

Chapter 26. Preliminary remarks on tectal polyhedra.

My original plan had been to treat tectal polyhedra in the same way as apical polyhedra, namely by classifying them by the number of sides of the polygons immediately adjacent to the roof. (Because we will have the occasion to use the phrase "face immediately adjacent to the roof" so frequently here, let us introduce a term: a *subtectal face* is a face of a tectal polyhedron which is immediately adjacent to its roof.) But further thought prompted me to revise this plan. There are two kinds of adjacency involved here: faces sharing a complete edge with the roof and those sharing only a vertex with the roof. There may be *no* faces that share only a vertex with the roof (this is the case for the prisms and prismoids, for example), while there *must* be (in any tectal polyhedron) faces sharing a complete edge with the roof; but in addition, the faces sharing a complete edge with the roof and those sharing only a vertex with the roof need not be of the same type of polygon, even if they both exist. So, if this were to be the basis for a classification of tectal polyhedra, the number of classes would proliferate fairly rapidly, as different numbers of sides of both kinds of subtectal faces were considered.

Instead, a different system of classification of the tectal polyhedra will be followed. One can see that if the principal rotation axis is n-fold, the roof and base must be n-gons, $2n$-gons, or in general kn-gons, where k is some integer. (If either is an n-gon, it must be regular; otherwise it will be, in the most general case, an $n \times k$-gon.) We will use the two values of k for the roof and base as the classification keys for tectal polyhedra. (The choice of which face of a tectal polyhedron is the "roof" and which is the "base" is arbitrary; in this book, it will be assumed that the one with the *smaller number of sides* is the roof if they differ; if they both have equal numbers of sides, the one which is *smaller* will be the roof; otherwise it does not matter, as those cases will generally have D_{nh} or S_{2nv} symmetry, and both of the faces in question will be identical anyway.) On a tectal polyhedron whose principal rotation axis is n-fold, a roof or base of which is an n-gon will be termed *unigeneral* (from "*genus*": there is one genus or type of edge to such a polygon); one which is a $2n$-gon (whether regular or an $n \times 2$-gon; because, even if it is a regular $2n$-gon, the sides fall into two different transitivity classes with respect to the entire polyhedron) will be termed *bigeneral*, and so forth. The polyhedra with both base and roof unigeneral will be termed *uni/unigeneral*, those with unigeneral roof and bigeneral base, *uni/bigeneral*, etc. (Note that, while this terminology would permit the use of such adjectives as *trigeneral*, *quadrigeneral*, etc., in this book, only *unigeneral* and *bigeneral* roofs and bases, and the tectal polyhedra with these two types of roofs and bases, will be considered. Anyone is welcome to go farther.)

Before getting to the various types of tectal polyhedra and their classification according to this scheme, it will be useful to introduce another definition: a *prismatoid* (note that this word has two letters not in the word *prismoid*) is a polyhedron whose vertices all lie in two parallel planes. It might be noted that if all but one of the vertices of a prismatoid are in one of those two planes, the polyhedron is a *pyramid*, and is not tectal, but *apicobasal*. However, in any other case, a prismatoid is a special case of a tectal polyhedron. Since, however, this book will primarily be concerned with polyhedra with axial symmetry, the only types of prismatoids with which we will be concerned will be the uni/unigeneral, uni/bigeneral, etc. polyhedra just defined. While not all of the uni/unigeneral, uni/bigeneral, etc. polyhedra are prismatoids, the simplest ones will be, and each of the chapters devoted to this class will begin by discussing the ones that are.

One thing that should also be noted is that if say, a pair of uni/bigeneral polyhedra with congruent bigeneral bases are fused (turning them base to base, of course!) a uni/unigeneral polyhedron is obtained. (It may not be convex, but will be if the dihedral angles at the base are small enough, of course.) In general, if an m-/n-general polyhedron is fused to an n-/p-general polyhedron, an m-/p-general polyhedron is obtained; however, only the specific case of uni/bigeneral polyhedra will be important for this book. The result of this is that there are uni/unigeneral polyhedra that cannot be sensibly described until the uni/bigeneral polyhedra which are fused to produce them have been discussed, and thus there are uni/unigeneral polyhedra which will not be discussed in Chapters 27 and 28, but will first be introduced in Chapters 29 and 30, with the uni/bigeneral polyhedra.

It should also be noted that from all the apical polyhedra discussed in previous chapters, a tectal polyhedron can be obtained by truncation – of the apex, if the polyhedron is apicobasal; of both the apex and antiapex, if it is diapical. (This will be described as *apical truncation* of the polyhedron.) There is also the possibility of creating a uni/bigeneral polyhedron from a uni/unigeneral polyhedron, by truncating all the vertices at the base, but not the roof. (This will be described as *basal truncation* of the polyhedron.) While a large number of new types of polyhedron can be obtained in this way, they will mostly not be described here because there are so many. However, it might be noted that the *same* type of polyhedron is obtained by apical truncation of a 5-pyramid and by basal truncation of a prismoid; whether one calls it an apically truncated 5-pyramid or a basally truncated prismoid is unimportant. In either case it has the same configuration: an n-gon for the roof, an $n\times2$-gon for the base, and n hexagons for lateral faces extending base-to-roof, with n triangles extending only part-way from the base toward the roof. (See Figure 51 below.) In fact, this type of polyhedron can be thought of in a third way: as a (fully) truncated pyramid. And in that context, it should be noted that if the principal rotation axis is threefold, and the truncation depths are properly set, this polyhedron is an Archimedean solid, the *truncated tetrahedron*.

Chapter 26. Preliminary remarks on tectal polyhedra.

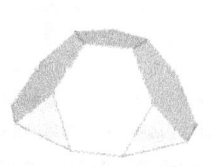

Figure 51: A truncated pyramid (or apically truncated 5-pyramoid, or basally truncated prismoid).

As was stated on page 108, if two uni/bigeneral polyhedra are fused together by their bases, the resulting polyhedron is a uni/unigeneral polyhedron. So as we will later see in Chapter 29, two cupolae or two rotundae can be fused together; the result, although not mentioned in Chapter 27, is a uni/unigeneral polyhedron, although its description has to be postponed until Chapter 29 because it requires the material in that chapter for easy discussion. But they can be fused in either of two ways: with the roofs of each in similar position, or half a rotation apart, giving (if the two rotundae or cupolae are congruent) respectively D_{nh} or S_{2nv} symmetry. (If they are not congruent, either position gives C_{nv} symmetry, but the two are still distinctly different.) It is to distinguish these that Johnson used the prefixes "ortho-" and "gyro-" that were introduced on p. 87 (Chapter 23) above. While he only referred to orthobicupolae and gyrobicupolae, it is convenient, as was stated on p. 87, to use them in any situation where this kind of alternative placement is possible. (A *"pentagonal gyrobirotunda"* is actually an Archimedean icosidodecahedron, if it is as regular as necessary to qualify as a Johnson solid.)

Chapter 27. Uni/unigeneral tectal polyhedra 1. General considerations.

The most frequently encountered of the uni/unigeneral tectal polyhedra are the prisms and antiprisms which were first reviewed in Chapter 5. The majority of this chapter will concern itself with these and their converged derivatives, the prismoid (which is obtained, as was already mentioned in Chapter 17, by converging a prism, and which was also introduced in Chapter 5) and the converged antiprism. All of these, of course, are special cases of prismatoids. Figure 53 depicts a prism, while Figure 57 depicts an antiprism.

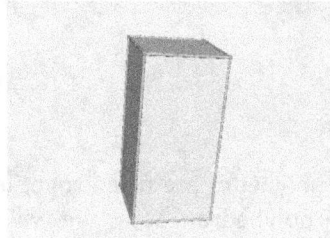

Figure 53: A square prism.

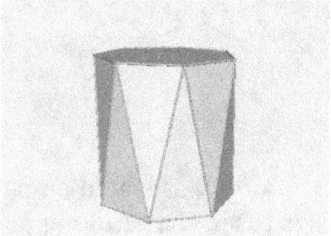

Figure 52: A heptagonal antiprism.

The prismoid, also known as a frustum of a pyramid, and the converged antiprism are depicted below in Figures 54 and 55.

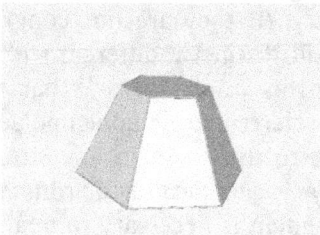

Figure 54: A hexagonal prismoid.

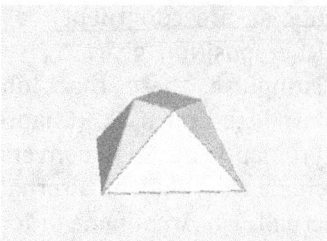

Figure 55: A square converged antiprism.

For all of these, one common description applies. If the polyhedron is to have n-fold symmetry (C_{nv}, S_{2nv}, or D_{nh}), the roof and base will both be regular n-gons. In the case of the prism and antiprism, they are congruent; in the case of the prismoid and converged antiprism, the roof is smaller than the base, but both are the same type of regular polygon. In contrast to this pairing, the prismoid and prism are alike in that both the roof and base are aligned the same way (corresponding sides parallel), while the other two polyhedra have roof and base aligned contrarily (maximum possible departure from parallelism: the sides of the two differ in direction by a turn of half as much as the n-fold rotation that is a symmetry transformation of the polyhedron). In the case of the prism and prismoid, vertices that are in corresponding position are connected by edges of the polyhedron; in the case of the antiprism and converged antiprism, each vertex of the roof is equidistant from two vertices of the base, and vice versa,

Chapter 27. Uni/unigeneral tectal polyhedra I. General considerations.

and edges connect it with both.

Leaving the category of prismatoids, one of the easiest types of uni/unigeneral tectal polyhedra to describe is the *rectified prism*, as illustrated in Figure 56. By referring, if necessary, to the definition of rectification in Chapter 9, one can see that a rectified regular *n*-gonal prism has D_{nh} symmetry, with two regular *n*-gons as roof and base, 2*n* subtectal triangles (at least isosceles, though if the dimensions of the bases and height are in the correct proportion, equilateral) from the truncation of the 2*n* vertices of the original prism, and *n* lateral faces that are at least rhombi, though if the dimensions of the bases and height are in the correct proportion, squares. When *n*=4, and if the starting prism is a cube, of course, the resulting polyhedron is a cuboctahedron.

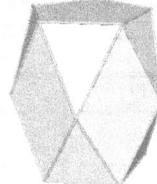

Figure 56: A rectified pentagonal prism.

The rectified prism, however, can be thought of one of nine different types of polyhedra obtained by truncation of a prism to different depths. When the process of truncation was explained in Chapter 9, there were three different intermediate stages of truncation of a cube, with a final, fourth, that was the dual octahedron. This final stage will be ignored here, because it is better done by one of the other dualization methods described in Chapter 12. In the case of the cube, in order to preserve the overall O_h symmetry, it was necessary to have the truncation equally deep along all edges meeting a vertex. But in a more general case of a prism, the lateral edges can be distinguished from the edges that are sides of the bases, and one can place the truncation planes in such a way as to cut to a shallow, critical, or deep extent independently on these two classes of edges. Thus the nine possibilities. If the truncation is shallow across both sets of edges, a truncated prism is made that is analogous to the truncated cube illustrated in Chapter 9. If it is critical across both sets of edges, a rectified prism such as is illustrated in Figure 56, analogous to the cuboctahedron, is produced. And if it is deep across both sets of edges, a truncated bipyramid, analogous to the truncated octahedron, is produced.

But there is no name appropriate to the six other cases, where the truncation of the edges that are sides of the bases is at a different level than the truncation of the lateral edges, so no such name can be inserted in Table 15 below. However, the table is included so that you can see the composition of the various truncations of the prism. Some are uni/unigeneral, some are bi/bigeneral (none are uni/bigeneral because we have assumed a symmetrical truncation at all $2n$ of the vertices). In every case, there are three transitivity classes of face, two faces (base and roof) which are either n-gons or $n \times 2$-gons, a set of $2n$ faces that start, in shallow truncation, as isosceles (or equilateral) triangles, but gain additional sides when the truncation planes that generate them begin to intersect, and finally the n lateral faces, which were rectangles (or squares) in the original prism and rhombi (or squares) in the rectified prism, but have additional sides when truncation is shallow along some edges. The exact number of sides of these depends on the depth of truncation, as shown in Table 15 below.

		Truncation of roof/base edges		
		Shallow	Critical	Deep
Truncation of lateral edges	Shallow	2 $n\times 2$-gons	2 n-gons	2 n-gons
		$2n$ triangles	$2n$ triangles	$2n$ pentagons
		n octagons	n hexagons	n hexagons
		Bi/bigeneral	Uni/unigeneral	Uni/unigeneral
		Truncated prism	no name	no name
	Critical	2 $n\times 2$-gons	2 n-gons	2 n-gons
		$2n$ triangles	$2n$ triangles	$2n$ pentagons
		n hexagons	n rhombi	n rhombi
		Bi/bigeneral	Uni/unigeneral	Uni/unigeneral
		no name	Rectified prism	no name
	Deep	2 $n\times 2$-gons	2 n-gons	2 n-gons
		$2n$ trapezoids	$2n$ trapezoids	$2n$ hexagons
		n hexagons	n rhombi	n rhombi
		Bi/bigeneral	Uni/unigeneral	Uni/unigeneral
		no name	no name	Truncated bipyramid

Table 15: The various polyhedra obtained by truncating a prism.

In order to understand how Table 15 was constructed, consider each set of faces separately. First: the roof and base: In the original prism, they are regular n-gons. As soon as the shallowest of the truncations along the roof/base edges is made, n additional sides are added, one at each original vertex, making them into $n\times 2$-gons. Each vertex of the original roof and base has been split into two, and one side of the $n\times 2$-gon (one of the "new" sides) joins the two vertices that had come from one original vertex, while another (one of the "original" sides) joins vertices that came from

Chapter 27. Uni/unigeneral tectal polyhedra I. General considerations.

originally adjacent vertices of the roof or base. As the truncation goes deeper, it passes from the case where the original sides are longer than the new ones to the case where the original sides are shorter than the new ones, with one depth at which the sides have equal length and they become regular $2n$-gons. When the truncation reaches the critical point, however, the original edges have been reduced to zero length. The polyhedron now has a new vertex, at what had been the midpoint of each side of the roof and base formed by merger of two vertices which had come from adjacent vertices of the original polyhedron. The roof and base have, at this point, been restored to n-gons, though rotated by $180°/n$ from their original position. Deep truncation only makes them smaller, though creating additional edges *crossing* the original roof/base edges, between the (originally triangular) face that had been created at each vertex. These, in the subsequent discussion, will be termed the "intermediate" edges; they are located halfway between what had been the original lateral edges of the prism. During all this, nothing depends on what was done to the lateral edges, as long as *some* truncation has occurred.

The lateral faces of the prism, as soon as any truncation takes place, become octagons instead of the rectangles they started as. In the most general case, they are 2×4-gons, as the two sides that come from lateral edges of the original prism need not be equal to the two that come from roof/lateral face and base/lateral face edges. If they are equal, the octagons are 4×2-gons, which can become regular octagons if the depth of the truncation reaches the point where the new edges produced by the truncation are equal to the sides remaining from the original prism. When the truncation becomes critical, two sides of the octagon disappear. If it is critical at the roof and base, the roof/lateral face and base/lateral face edges become zero; if it is critical on the lateral edges, they are the ones that become zero. If either set alone becomes zero, the octagon becomes a hexagon (and it can be seen, the two cases orient the hexagons differently, depending on whether it is the roof/lateral face and base/lateral face edges or the lateral edges which have some parts remaining. Going from critical to deep truncation, those lost edges remain and the hexagons only become smaller. When both truncations become critical and/or deep, however, both sets of lost edges are involved and the lateral faces have become rhombi.

The new faces created in the truncation planes show the most complicated behavior. In the shallowest truncations, they are triangular. They remain so as long as the truncations are no deeper than critical. If the roof/base truncation becomes deep, two additional sides are introduced, the intermediate edges of the polyhedron discussed earlier. If the lateral truncation becomes deep, one additional side is introduced, as an edge halfway between the roof and base develops. Since either of these can happen separately, the triangle can become a pentagon (if only the roof/base truncation is deep), a trapezoid (if only the lateral truncation is deep) or a hexagon (if both are).

Because of the original D_{nh} symmetry of the prism, requiring among other things that mirror planes pass through the lateral edges, these polygons (at the original prism vertices) are in all cases at least isosceles. (They may have more symmetry than that, depending on the depth of the truncation and the original edge lengths.)

There is a class of uni/unigeneral tectal polyhedra that will not be discussed in this chapter but will be taken up in Chapters 29 and 30. The reason for this is that their construction is best described by the fusion of two uni/bigeneral tectal polyhedra, as mentioned on p. 108 (Chapter 26), and therefore their introduction is best postponed until the uni/bigeneral tectal polyhedra have been introduced.

Chapter 28. Uni/unigeneral tectal polyhedra II. Mathematical details.
[Optional]

In this chapter, exactly as in the previous chapter, the majority of the emphasis will be placed on the four uni/unigeneral tectal polyhedra which are prismatoids and possess mirror planes through the principal rotation axis. As *uni/unigeneral* tectal polyhedra, they must have a base and roof which are regular *n*-gons, where *n* is the order of the principal rotation axis of the polyhedron. These polyhedra can be grouped as two that have *aligned* roof and base (the prism and prismoid), and two that have *staggered* roof and base (the antiprism and converged antiprism). Alternatively, they can be grouped as two that have *congruent* roof and base (the prism and antiprism), and two that have *similar, noncongruent* roof and base (the prismoid and converged antiprism). For all these four, the coordinates of the 2n different vertices are easy to represent. Using the usual conventions of this book, the coordinates of the roof will be designated as A_1, A_2, ..., A_n; and the coordinates of the base will be designated as B_1, B_2, ..., B_n. Then the simplest case is that of the prism, which will be represented with the origin at the centroid of the polyhedron:

A_j: $(r \cos [j-1]\alpha, r \sin [j-1]\alpha, z_A)$,

B_j: $(r \cos [j-1]\alpha, r \sin [j-1]\alpha, -z_A)$,

where $\alpha = 2\pi/n = 360°/n$, and in each case, $j = 1, 2, ..., n$. For the antiprism, the only difference is that, in order to stagger the vertices, the base is rotated by $\alpha/2$, leading to the coordinates:

A_j: $(r \cos [j-1]\alpha, r \sin [j-1]\alpha, z_A)$,

B_j: $(r \cos [j-½]\alpha, r \sin [j-½]\alpha, -z_A)$.

For the two other polyhedra, the only difference is that a single value of *r* cannot be used for both the A_j and the B_j, so one must write for the prismoid:

A_j: $(r_A \cos [j-1]\alpha, r_A \sin [j-1]\alpha, z_A)$,

B_j: $(r_B \cos [j-1]\alpha, r_B \sin [j-1]\alpha, z_B)$,

though, since the centroid is no longer going to be halfway between the roof plane and the base plane, it is better to write z_A and z_B for the z-coordinates (as has been done here) than the z_A and $-z_A$ that were written for the prism. And similarly, the converged antiprism can be given the coordinates:

A_j: $(r_A \cos [j-1]\alpha, r_A \sin [j-1]\alpha, z_A)$,

B_j: $(r_B \cos [j-½]\alpha, r_B \sin [j-½]\alpha, z_B)$.

The only other polyhedra whose coordinates will be taken up in this chapter will be the *rectified* prism and prismoid; rectifying the antiprism and converged antiprism is certainly possible, but the resulting polyhedra are difficult to visualize and, in this author's opinion, rather devoid of the beauty of the highly symmetrical polyhedra with which this book has mostly been concerned. Since the rectified prism and prismoid are not prismatoids, the coordinates of the vertices on the roof and base are not to be represented here by $A_1, A_2, ..., A_n$ and $B_1, B_2, ..., B_n$, but as $A_1, A_2, ..., A_n$ and $C_1, C_2, ..., C_n$, in order to allow for the intermediate vertex rotation plane to use $B_1, B_2, ..., B_n$. But the actual coordinates are built up the same way, only replacing B by C when it occurs. In order to distinguish the vertices of the *original* prism from those of the *rectified* prism, the value of r used to determine the coordinates of the original prism will be designated r_p, and the vertices by A_{pj} and B_{pj}, so that one can write:

A_{pj}: $(r_p \cos [j-1]\alpha, r_p \sin [j-1]\alpha, z_A)$,

B_{pj}: $(r_p \cos [j-1]\alpha, r_p \sin [j-1]\alpha, -z_A)$

Then for the rectified prism, it is clear that the coordinates of A_j are halfway between those of A_{pj} and $A_{p(j+1)}$, so the x-coordinate is

$$r_p[\cos j\alpha + \cos (j-1)\alpha]/2$$

$$= r_p[\cos (j-½)\alpha \cos \alpha/2 - \sin (j-½)\alpha \sin \alpha/2 + \cos (j-½)\alpha \cos \alpha/2 + \sin (j-½)\alpha \sin \alpha/2]/2$$

$$= r_p[\cos (j-½)\alpha \cos \alpha/2],$$

and the y-coordinate is

$$r_p[\sin j\alpha + \sin (j-1)\alpha]/2$$

$$= r_p[\sin (j-½)\alpha \cos \alpha/2 + \cos (j-½)\alpha \sin \alpha/2 + \sin (j-½)\alpha \cos \alpha/2 - \cos (j-½)\alpha \sin \alpha/2]/2$$

$$= r_p[\sin (j-½)\alpha \cos \alpha/2].$$

Rotating the entire prism by $\alpha/2$, the coordinates can be written with $(j-½)\alpha$ replaced by $(j-1)\alpha$; and letting $r = r_p \cos \alpha/2$, for the rectified prism one can write:

A_j: $(r \cos [j-1]\alpha, r \sin [j-1]\alpha, z_A)$,

C_j: $(r \cos [j-1]\alpha, r \sin [j-1]\alpha, -z_A)$.

The vertices in the B plane are similarly located by putting B_j halfway between A_{pj} and B_{pj}, and since we had set $r = r_p \cos \alpha/2$, we can write $r_p = r/\cos \alpha/2$; because we had rotated the prism by $\alpha/2$ in generating the coordinates of the A_j and C_j, we must allow for this in writing

B_j: $(r \{\cos [j-½]\alpha\}/\cos [\alpha/2], r \{\sin [j-½]\alpha\}/\cos [\alpha/2], 0)$.

It is clear that the same procedure could be followed to determine the coordinates of the vertices of a rectified prismoid starting with the coordinates given above for a prismoid; however, a different procedure will be shown, just to demonstrate an alternative way of accomplishing the same result. It is clear that symmetry requirements dictate that the three sets of coordinates be given by:

A_j: $(r_A \cos [j-1]\alpha, r_A \sin [j-1]\alpha, z_A)$,

B_j: $(r_B \cos [j-½]\alpha, r_B \sin [j-½]\alpha, z_B)$,

C_j: $(r_C \cos [j-1]\alpha, r_C \sin [j-1]\alpha, z_C)$,

with $z_C < z_B < z_A$. It will be convenient to take $z_B = 0$, and simply require $z_C < 0$ and $z_A > 0$, though one cannot assume that $z_C = -z_A$ in this case, as for the prism. The only condition that can be imposed here is that A_1, B_n, B_1, and C_1 are all in one plane. (By applying rotational symmetry, all the other faces will be planar as needed.) Explicitly,

A_1: $(r_A, 0, z_A)$,

B_n: $(r_B \cos \alpha/2, -r_B \sin \alpha/2, 0)$,

B_1: $(r_B \cos \alpha/2, r_B \sin \alpha/2, 0)$, and

C_1: $(r_C, 0, z_C)$

must all be in one plane. This will be ensured by making certain that the midpoint of the line segment B_1B_n lies on the segment A_1C_1. Now the midpoint of the segment B_1B_n is easily seen to be given by $(r_B \cos \alpha/2, 0, 0)$. And the segment A_1C_1 crosses the xy-plane at $(r, 0, 0)$ where

$$r = (r_A z_C + r_C z_A)/(z_C + z_A).$$

These will be the same point if

$$r_B = (r_A z_C + r_C z_A)/(z_C + z_A)\cos \alpha/2.$$

So with that condition imposed on r_B, the coordinates

A_j: $(r_A \cos [j-1]\alpha, r_A \sin [j-1]\alpha, z_A)$,

B_j: $(r_B \cos [j-½]\alpha, r_B \sin [j-½]\alpha, 0)$,

C_j: $(r_C \cos [j-1]\alpha, r_C \sin [j-1]\alpha, z_C)$,

define a rectified prismoid.

Chapter 29. Uni/bigeneral tectal polyhedra 1. General considerations.

The monocropped antiprism, mentioned in Chapter 17, is an example of a uni/bigeneral tectal polyhedron. In Norman Johnson's treatment of the solids which now bear his name, he introduced two terms for particular solids which also fall within the category we count as uni/bigeneral tectal polyhedra: the *cupola* and *rotunda*. However, while Johnson's cupolae and rotundae were restricted to those which could be constructed with only regular polygons, we will use the term for all those which fall under a more general definition, which include his as special cases, and which will be discussed in this chapter. Certainly, at least for the cupolae, the generalization proposed in here is not new to this book (it can be found in the online encyclopedia *Wikipedia*, for example).

The monocropped antiprism and the cupola are, in a number of ways, quite similar, and also, because they are prismatoids, are the easiest uni/bigeneral tectal polyhedra to describe, so this chapter will begin with them.

First, consider a prismatoid which is to satisfy the definition of a uni/bigeneral tectal polyhedron. It must, by the definition on page 107 in Chapter 26, have an n-fold principal rotation axis. Since it is a prismatoid, all the vertices must lie in two planes; since it is a uni/bigeneral tectal polyhedron, n of these vertices must lie in one plane, which we take to be the roof according to the convention in the same paragraph on page 107, and $2n$ of these vertices must lie in the other plane, which we take to be the base according to the same convention. The main choices we have at this point is how these vertices are arranged in those planes, and the existence of an n-fold principal rotation axis places limits on this as well. On the roof, the n vertices must form a regular n-gon. On the base, at the very minimum they must form an $n \times 2$-gon; they can, of course, form a regular $2n$-gon. First suppose that they actually do form a regular $2n$-gon. If the sides are all equal to the sides of the roof n-gon, the presence of the n additional sides means the base polygon will extend further from the axis than the roof polygon, and it would be a particularly symmetrical arrangement to make *alternate* sides of the base parallel to *all* the sides of the roof. In that case, making an edge of the polyhedron by connecting each vertex of the base to that vertex of the roof nearest to it means that each vertex of the roof is connected to two vertices of the base polygon. These edges from any roof vertex to two vertices of the base form, together with the edge joining those two vertices, a triangle which, at the very least, is isosceles, and can be, if the distance between the planes has the proper relationship to the sides of the base and roof polygons, equilateral. These isosceles or equilateral triangular faces can be referred to as *lateral* triangular faces. If the remaining faces (besides the base, the roof, and the lateral triangular faces), are now considered, they can be seen to be quadrilaterals in which two sides were made equal and parallel, implying at least that they are parallelograms, but in fact, because of the overall C_{nv} symmetry, they must be at least *rectangles*, and again can be, if the distance between the planes has the proper relationship to the sides of the base and roof polygons, squares. If the triangles are in fact equilateral, the rectangles will also be squares, since the equilaterality of the triangles implies

Chapter 29. Uni/bigeneral tectal polyhedra I. General considerations.

that all the lateral edges are equal in length to the edges of the base and roof polygons. And this sort of polyhedron, with a regular *n*-gonal roof, a regular 2*n*-gonal base, and lateral faces consisting of *n* equilateral triangles and *n* squares, is what Norman Johnson called a *cupola*. An example is shown below in Figure 57.

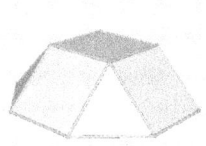

Figure 57: A Johnson square cupola.

Because two squares and an equilateral triangle have a total angle sum at a roof vertex of 240°, the *n*-gon at the roof cannot have an angle of 120° or greater, thus the Johnson cupolae are restricted to *n*=3, 4, and 5. (See Table 16.) But in this book, as in *Wikipedia*, however, the term *cupola* will be used for the polyhedron even when the triangles are isosceles and the rectangles are not square. And in fact, even when the base is not a regular 2*n*-gon, but only an *n*×2-gon, if alternate sides are equal and parallel to the sides of the roof, so that the lateral quadrilateral faces are rectangular, the polyhedron will be considered to be designated as a cupola in this book. As we have done in the case of terms like "Archimedean truncated octahedron," where the term "truncated octahedron" is used more broadly in this book than it is often used, so that the adjective "Archimedean" is added when the narrower sense is intended, the use of the term "Johnson" will be used in this book when it is intended to restrict the use of such names as "square cupola" to the ones that are Johnson solids. See, for example, Figure 58 below.

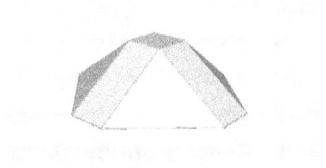

Figure 58: A square cupola, more generally defined.

We have described these cupolae in terms of a regular *n*-gonal roof, an *n*×2-gonal base, and 2*n* lateral faces, *n* of which are isosceles triangles and the remaining *n* rectangles. The three Johnson cupolae are special cases where the base is a regular 2*n*-gon and the lateral faces are equilateral triangles and squares. If one takes this recipe for a polygon of C_{nv} symmetry and reduces *n* to 2, the "polygon" at the roof becomes a straight line segment. If one calls this a "digon" as was done with the antiprism (see p. 27, Chapter 5), the cupola becomes a triangular prism (with *one lateral face* of the *prism* considered as the *base* of the *cupola*, and the *bases* of the *prism* considered as *lateral faces* of the *cupola*). It is not really a new polygon, but it is another case of the same polyhedron being generated by two different processes. While we would not normally think of the *"digonal cupola"* as a different thing from a triangular prism, it is a full-fledged member of the family of cupolae, just as a cube is both a square prism and a triangular antibipyramid.

For all of these cupolae, whether defined in Johnson's narrower definition or the broader one here, the base is necessarily larger than the roof because half of its sides are equal to the sides of the roof polygon, and there are other sides between them. If the sides of the roof are *larger* than the parallel sides of the base, however, the lateral quadrilateral faces become isosceles trapezoids rather than rectangles. If the difference in length between the sides of the roof and the parallel sides of the base is great enough, the edges between these isosceles trapezoids will converge, rather than diverging away from the principal axis of rotation; the resulting polyhedron is a monocropped antiprism. An example is shown in Figure 59 below. (Note that Figure 59 is identical to Figure 35 in Chapter 16, but with the top-to-bottom direction reversed.)

Figure 59: An example of a monocropped hexagonal antiprism.

It is certainly possible to produce a uni/bigeneral tectal polyhedron that is not a prismatoid. This only requires the addition of at least one set of vertices in a plane between the planes of the roof and base polygons. If there is only one such plane, of course, the polyhedron may be a composite polyhedron, as we have defined it in

Chapter 29. Uni/bigeneral tectal polyhedra I. General considerations.

Chapter 14 (p. 63), and though we are not much interested in composite polyhedra, some of them (particularly the elongated and gyroelongated cupolae) qualify as Johnson solids and will be included when the Johnson solids that fall into our classifications are enumerated. However, it is also possible that a polyhedron with only one intermediate vertex rotation plane can be a simple polyhedron that falls into the category of uni/bigeneral tectal polyhedra. The only one that Norman Johnson considered, because it can be constructed with regular pentagons and equilateral triangles for lateral faces, is the one he named the rotunda. The only rotunda that Johnson described was the *pentagonal* rotunda, because it is the only one that can be made with all regular polygons; once more, as with the cupola, in this book the definition will be extended.

For the purposes of this book, a rotunda will be defined as follows: A polyhedron with C_{nv} symmetry with a roof that is an *n*-gon and a base that is an *n*×2-gon, with alternate sides of the base parallel to the sides of the roof, containing *n* vertices in a plane between the roof and base plane, each located in the plane that bisects a side of the roof and the parallel side of the base; the edges of the polyhedron are the sides of the roof, the sides of the base, and lines connecting the intermediate vertices to the ends of the nearest sides of the roof and base.

Not mentioned in this definition are the pentagonal and triangular faces created by making the connections described between the intermediate vertices and the ends of the nearest sides of the roof and base. It is not necessary to mention them in the definition, because they are produced automatically by constructing a rotunda according to the definition. Figure 60 shows a square rotunda, not a Johnson solid because the triangles and pentagons are not regular.

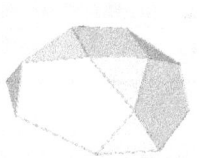

Figure 60: A square rotunda.

It is impossible, as stated earlier, to make a rotunda that is a Johnson solid unless the roof is a pentagon; it seems rather odd, therefore, that Johnson defined a new term, appropriate to a whole family of polyhedra, that only contained one member that fit his criteria. But perhaps, because he could also include elongated and gyroelongated pentagonal rotundae, he felt the term useful. Of course, by expanding its usage, it becomes a truly useful term to describe a whole family of uni/bigeneral tectal polyhedra.

Another instance, this one not named by Johnson, of a non-prismatoidal uni/bigeneral tectal polyhedron is the truncated pyramid described on p. 108 in Chapter 26 and illustrated in Figure 51 on p. 109. Although the family of truncated pyramids (or apically truncated 5-pyramoids, or basally truncated prismoids) includes one Archimedean solid (the truncated tetrahedron), it is apparently not possible to make all the polygons regular when the principal rotation axis is other than threefold, and therefore none of its members are Johnson solids. But it belongs to the larger family of uni/bigeneral tectal polyhedra, and so is included in this chapter.

For easy reference, the examples of uni/bigeneral tectal polyhedra covered in this chapter which also qualify as Johnson solids are given in Table 16 below. Of course, the elongated and gyroelongated cupolae and rotunda are *composite* polyhedra according to the definition on p. 63 in Chapter 14, and not of as great interest as the simple ones, but because of their inclusion as Johnson solids are important for classification purposes.

Name	Designation as a Johnson solid
Triangular cupola	J_3
Square cupola	J_4
Pentagonal cupola	J_5
Elongated triangular cupola	J_{18}
Elongated square cupola	J_{19}
Elongated pentagonal cupola	J_{20}
Gyroelongated triangular cupola	J_{22}
Gyroelongated square cupola	J_{23}
Gyroelongated pentagonal cupola	J_{24}
Pentagonal rotunda	J_6
Elongated pentagonal rotunda	J_{21}
Gyroelongated pentagonal rotunda	J_{25}

Table 16: Uni/bigeneral tectal polyhedra which also qualify as Johnson solids.

As stated above, a "digonal cupola" is the same as a triangular prism, and is not a Johnson solid; however, it can take part in some of the fusions that the other cupolae can, including that of two of them, base to base. Johnson did not call a digonal gyrobicupola by that name – he never used the term *"digonal"* – he instead called it a gyrobifastigium, a name we mention here for completeness, but will not use elsewhere in this book. Although they are uni/unigeneral, the polyhedra obtained by base-to-base

Chapter 29. Uni/bigeneral tectal polyhedra I. General considerations.

fusion of two of the polyhedra of Table 16 could not sensibly be discussed in previous chapters, since their only convenient description is in terms of fusions of the polyhedra introduced in this chapter. As was the case in the naming of the bipyramid on p. 71, Chapter 19, when two cupolae or two rotundae are fused at their bases, the term "bicupola" or "birotunda" is used to designate the resulting polyhedron. However, unlike the case of pyramids, there are two ways a pair of cupolae or rotundae can be fused: so as to make base and roof similarly aligned (with all sides of the base parallel to sides of the roof) or oppositely aligned (rotated $180°/n$ around the principal rotation axis with respect to each other). In the first case, the prefix "ortho-" is used; in the second "gyro-" in Johnson's nomenclature. A similar use of "ortho-" and "gyro-" is used in the fusion of a cupola to a rotunda to produce a "cupolarotunda." These polyhedra could not be mentioned in Chapter 27, because the uni/bigeneral polyhedra from which they are obtained by fusion were not named until this chapter, so they are therefore listed below in Table 17.

Name	Designation as a Johnson solid
Triangular orthobicupola	J_{27}
Square orthobicupola	J_{28}
Pentagonal orthobicupola	J_{30}
Elongated triangular orthobicupola	J_{35}
Elongated pentagonal orthobicupola	J_{38}
Digonal gyrobicupola ("gyrobifastigium")	J_{26}
Square gyrobicupola	J_{29}
Pentagonal gyrobicupola	J_{31}
Elongated triangular gyrobicupola	J_{36}
Elongated square gyrobicupola	J_{37}
Elongated pentagonal gyrobicupola	J_{39}
Gyroelongated triangular bicupola	J_{44}
Gyroelongated square bicupola	J_{45}

Name	Designation as a Johnson solid
Triangular orthobicupola	J_{27}
Gyroelongated pentagonal bicupola	J_{46}
Pentagonal orthocupolarotunda	J_{32}
Elongated pentagonal orthocupolarotunda	J_{40}
Pentagonal gyrocupolarotunda	J_{33}
Elongated pentagonal gyrocupolarotunda	J_{41}
Gyroelongated pentagonal cupolarotunda	J_{47}
Pentagonal orthobirotunda	J_{34}
Elongated pentagonal orthobirotunda	J_{42}
Elongated pentagonal gyrobirotunda	J_{43}
Gyroelongated pentagonal birotunda	J_{48}

Table 17: Uni/unigeneral tectal polyhedra, formed from the fusion of uni/bigeneral tectal polyhedra, which also qualify as Johnson solids.

Note that a triangular gyrobicupola is not listed because a Johnson triangular gyrobicupola would be a cuboctahedron. Similarly, there is no listing of an elongated square orthobicupola, because that would be a (small) rhombicuboctahedron. And it was already noted that a pentagonal gyrobirotunda would be an icosidodecahedron.

Chapter 30. Uni/bigeneral tectal polyhedra II. Mathematical details.
[Optional]

The only uni/bigeneral tectal polyhedra with which this chapter will be concerned are the ones that are *prismatoids;* i. e., polyhedra whose vertices lie in two parallel planes. The symmetry of these uni/bigeneral tectal polyhedra will be C_{nv}, for some value of n. Consequently, the vertices in one plane will form a regular n-gon, while the vertices in the other will either form a regular $2n$-gon or, more generally, an $n\times 2$-gon. It will be convenient to designate the vertices of the regular n-gon as A_j (j = 1, 2, ..., n) and those of the $n\times 2$-gon as B_j (j = 1, 2, ..., $2n$). With the z-axis oriented along the C_n axis of the polyhedron, and a suitable choice of the orientation of the x and y-axes, the coordinates of the vertices become:

A_j: ($r_A \cos [j - 1]\alpha$, $r_A \sin [j - 1]\alpha$, z_A), where α = $2\pi/n$ = $360°/n$,

B_j: ($r_B \cos \{[i - 1]\alpha + \beta\}$, $r_B \sin \{[i - 1]\alpha + \beta\}$, z_B), when j = $2i - 1$,

B_j: ($r_B \cos [i\alpha - \beta]$, $r_B \sin [i\alpha - \beta]$, z_B), when j = $2i$.

Because of the symmetry of the polyhedron, the only lateral faces that are needed to consider are the triangle $A_1B_1B_{2n}$ and quadrilateral $A_1A_2B_2B_1$. For these, the coordinates are:

A_1: (r_A, 0, z_A),

A_2: ($r_A \cos \alpha$, $r_A \sin \alpha$, z_A),

B_1: ($r_B \cos \beta$, $r_B \sin \beta$, z_B),

B_2: ($r_B \cos [\alpha - \beta]$, $r_B \sin [\alpha - \beta]$, z_B),

B_{2n}: ($r_B \cos \beta$, $-r_A \sin \beta$, z_B).

For the last of these, one should note that

$$\cos [n\alpha - \beta] = \cos [2\pi - \beta] = \cos \beta, \text{ and}$$
$$\sin [n\alpha - \beta] = \sin [2\pi - \beta] = -\sin \beta,$$

because of the definition of α = $2\pi/n$. Since any three points are automatically coplanar, it is clear that no special computation is needed to assure that $A_1B_1B_{2n}$ is a plane triangle. But it must be *demonstrated* that $A_1A_2B_2B_1$ is a plane quadrilateral. To show that, it is necessary simply to show that the vectors A_1A_2 and B_1B_2 are parallel. But this can be proved by using the *two-plane parallel rule* demonstrated in Appendix B (p. 138). In the notation of the rule as it is there, the point P_1 corresponds to A_1 here, P_2 corresponds to A_2, P_3 corresponds to B_1, and P_4 corresponds to B_2. The angle θ is 0, ϕ is α, and β has the same significance it does here.

But this means that A_1A_2 and B_1B_2 are parallel. And if A_1A_2 and B_1B_2 are parallel, then this means that A_1, A_2, B_1, and B_2 are all in one plane, so $A_1A_2B_2B_1$ is a plane quadrilateral. (In fact, it must be either a rectangle – possibly even a square – or an isosceles trapezoid, because of the C_{nv} symmetry of the polyhedron.)

The specific nature of this polyhedron depends on the values of r_A, r_B, n, and β. If $A_1A_2B_2B_1$ is a rectangle (square or not) then the polyhedron is a cupola. (It would have to be square, *and* the triangle $A_1B_1B_{2n}$ would have to be equilateral, for the polyhedron to be a *Johnson* cupola.) If $A_1A_2B_2B_1$ is an isosceles trapezoid, and $B_1B_2 < A_1A_2$, then A_1B_1 and A_2B_2 would, if extended, meet at some point beyond the B plane. If we call this point C_1, and locate the various C_j in corresponding positions by symmetry (i. e., C_j is the intersection point of A_jB_i, extended, with $A_{j+1}B_{i+1}$, extended, where $j = 2i - 1$), then the A_j and C_j points are the vertices of either an antiprism or a converged antiprism, depending on the value of the radius of the circle through the C_j points. This would make the original polyhedron whose vertices are the A_j and B_j points either a monocropped antiprism or a monocropped converged antiprism.

Chapter 31. Bi/bigeneral tectal polyhedra 1. General considerations.

The only type of bi/bigeneral tectal polyhedron which we will consider in this book is the *quasiprism* which has been mentioned earlier in different contexts. In Chapter 16 the quasiprism was described as a polyhedron obtained by the parallel cropping of both bases of an antiprism. In this chapter a different definition will be used, which does not appear to be the same, but it will be shown in the following chapter that they are equivalent, the demonstration being reserved to an optional chapter because the mathematics is relatively difficult.

Consider two $n\times 2$-gons in two parallel planes, with their centers directly in line with each other on an axis perpendicular to those planes. To form a quasiprism with an n-fold axis of rotation (actually a $2n$-fold alternating axis) the longer set of the sides of one must be oriented parallel to the shorter set of the sides of the other, and vice versa. In this book, the term "quasiprism" will normally be used specifically to describe a polyhedron in which the two $n\times 2$-gons are congruent: the longer sides of one are equal in length to those of the other, as are the shorter sides. It might be appropriate to consider a less-symmetric arrangement in which the sides of one and those of the other have different lengths, but even in that case one would insist that the longer sides of one are oriented parallel to the shorter set of the sides of the other, and vice versa, as stated earlier in this paragraph. The lateral faces of the quasiprism are isosceles trapezoids, such that the parallel sides of the trapezoids are parallel sides of the two $n\times 2$-gons, and the other sides of the trapezoids are lines joining corresponding vertices of those sides. While a quasiprism was illustrated in Figure 36 (on p. 67 in Chapter 16), another more easily recognized view is given below in Figure 61.

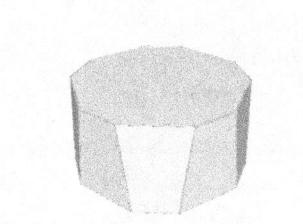

Figure 61: A pentagonal quasiprism.

Chapter 32. Bi/bigeneral tectal polyhedra II. Mathematical details.
[Optional]

In Chapter 16, a quasiprism was described as a polyhedron obtained by the parallel cropping of both bases of an antiprism. In Chapter 31, a quasiprism with an n-fold axis of rotation (actually a $2n$-fold alternating axis) is described as a polyhedron with two $n\times2$-gonal faces in parallel planes, with their centers directly in line with each other on an axis perpendicular to those planes, having the longer set of the sides of one $n\times2$-gonal face oriented parallel to the shorter set of the sides of the other, and vice versa. The primary goal of this chapter is to show how one might demonstrate the equivalence of these two definitions. The full demonstration involves such a long algebraic argument that only the beginning is given here; even this beginning of a demonstration runs so many pages that it warrants a chapter of its own.

For a complete demonstration of the equivalence of these two definitions, it is necessary to demonstrate both that any polyhedron meeting the description of a quasiprism in Chapter 16 also satisfies the definition in Chapter 31 and vice versa. The easier task is to show that a bicropped antiprism always has the properties of a quasiprism described in Chapter 31, and only that portion of the demonstration will be begun.

It might be thought desirable to define a quasiprism in a more general manner: a tectal polyhedron with a base and roof that are both polygons of the same number of sides, with corresponding sides of each parallel, and lateral faces which are quadrilaterals containing opposite sides which are corresponding sides of the roof and base. Such a generalized quasiprism would have, like the ones defined in Chapters 16 and 31, a base and roof which are polygons of the same type (though not congruent) and isosceles trapezoidal lateral faces, equal in number to the number of sides of the base and roof polygons, but would not have the S_{nv} symmetry of the quasiprisms as defined earlier, nor would it be properly described as a bi/bigeneral polyhedron, so at least for the purpose of this book, the term "quasiprism" will be restricted to the type of polyhedron defined in Chapters 16 and 31.

Consider an antiprism with the vertices of one base labeled P_1, P_2, ..., P_n, and the vertices of the other base labeled Q_1, Q_2, ..., Q_n. If we fix the z-axis along the line through the centers of the two bases, with the origin halfway between them, the coordinates of these points can be determined as:

$$P_k: (r\cos[k-1]\alpha,\ r\sin[k-1]\alpha,\ z_P),$$

$$Q_k: (r\cos[k-½]\alpha,\ r\sin[k-½]\alpha,\ -z_P)$$

with $k = 1, 2, ..., n$, and $\alpha = 360°/n$. The bicropping of the antiprism creates $2n$ vertices on each of the cropping planes, each P_k being replaced by two vertices A_{2k-1} and A_{2k}, and each Q_k being replaced by two vertices B_{2k} and B_{2k+1}. To simplify the algebra, we will write

$$x_{P_k} = r\cos[k-1]\alpha,$$

Chapter 32. Bi/bigeneral tectal polyhedra II. Mathematical details. [Optional]

$$y_{Pk} = r \sin [k - 1]\alpha,$$

$$x_{Qk} = r \cos [k - \tfrac{1}{2}]\alpha,$$

$$y_{Qk} = r \sin [k - \tfrac{1}{2}]\alpha.$$

The two planes by which the antiprism is cropped will be assumed to be symmetrically disposed about the origin, so their equations can be written as $z = \pm f z_P$. The several edges $A_{2k-1} B_{2k-1}$ are derived from the original edges $P_k Q_k$, while the edges $A_{2k} B_{2k}$ are derived from the original edges $P_{k+1} Q_k$, so the coordinates of all the points A_{2k-1}, B_{2k-1}, A_{2k}, and B_{2k} can be determined by where the planes $z = \pm f z_P$ cut the edges $P_k Q_k$ and $P_{k+1} Q_k$. Using the vector methods of Appendix B, it is necessary to find

$$x_{A(2k-1)} = (1 - g)x_{Pk} + g x_{Qk},$$

$$y_{A(2k-1)} = (1 - g)y_{Pk} + g y_{Qk},$$

$$z_{A(2k-1)} = (1 - g)z_P + g(-z_P) = (1 - 2g)z_P,$$

$$x_{B(2k-1)} = g x_{Pk} + (1 - g)x_{Qk},$$

$$y_{B(2k-1)} = g y_{Pk} + (1 - g)y_{Qk},$$

$$z_{B(2k-1)} = g z_P + (1 - g)(-z_P) = (2g - 1)z_P,$$

$$x_{A(2k)} = (1 - g)x_{P(k+1)} + g x_{Qk},$$

$$y_{A(2k)} = (1 - g)y_{P(k+1)} + g y_{Qk},$$

$$z_{A(2k)} = (1 - g)z_P + g(-z_P) = (1 - 2g)z_P,$$

$$x_{B(2k)} = g x_{P(k+1)} + (1 - g)x_{Qk},$$

$$y_{B(2k)} = g y_{P(k+1)} + (1 - g)y_{Qk}, \text{ and}$$

$$z_{B(2k)} = g z_P + (1 - g)(-z_P) = (2g - 1)z_P.$$

But it has already been determined that $z_A = f z_P$ and $z_B = -f z_P$ for all the points A_j and B_j, whether $j = 2k$ or $j = 2k - 1$, which requires

$$1 - 2g = f$$

or in other terms,

$$g = (1 - f)/2,$$

$$1 - g = (1 + f)/2.$$

Combining these with the equations for $x_{A(2k-1)}$, $y_{A(2k-1)}$, $x_{B(2k-1)}$, $y_{B(2k-1)}$, $x_{A(2k)}$, $y_{A(2k)}$, $x_{B(2k)}$, and $y_{B(2k)}$ gives

$$x_{A(2k-1)} = [(1+f)x_{Pk} + (1-f)x_{Qk}]/2,$$

$$y_{A(2k-1)} = [(1+f)y_{Pk} + (1-f)y_{Qk}]/2,$$

$$x_{B(2k-1)} = [(1-f)x_{Pk} + (1+f)x_{Qk}]/2,$$

$$y_{B(2k-1)} = [(1-f)y_{Pk} + (1+f)y_{Qk}]/2,$$

$$x_{A(2k)} = [(1+f)x_{P(k+1)} + (1-f)x_{Qk}]/2,$$

$$y_{A(2k)} = [(1+f)y_{P(k+1)} + (1-f)y_{Qk}]/2,$$

$$x_{B(2k)} = [(1-f)x_{P(k+1)} + (1+f)x_{Qk}]/2, \text{ and}$$

$$y_{B(2k)} = [(1-f)y_{P(k+1)} + (1+f)y_{Qk}]/2.$$

But the various values of x_{Pk}, x_{Qk}, y_{Pk}, and y_{Qk} were already defined (and, of course, the corresponding ones for k replaced by $k+1$ can be expressed the same way by replacing k by $k+1$ in the equations) as

$$x_{Pk} = r \cos (k-1)\alpha,$$

$$y_{Pk} = r \sin (k-1)\alpha,$$

$$x_{Qk} = r \cos (k-\tfrac{1}{2})\alpha, \text{ and}$$

$$y_{Qk} = r \sin (k-\tfrac{1}{2})\alpha.$$

Substituting these in the above equations for $x_{A(2k-1)}$, $y_{A(2k-1)}$, $x_{B(2k-1)}$, $y_{B(2k-1)}$, $x_{A(2k)}$, $y_{A(2k)}$, $x_{B(2k)}$, and $y_{B(2k)}$ gives

$$x_{A(2k-1)} = [(1+f)\cos(k-1)\alpha + (1-f)\cos(k-\tfrac{1}{2})\alpha]r/2,$$

$$y_{A(2k-1)} = [(1+f)\sin(k-1)\alpha + (1-f)\sin(k-\tfrac{1}{2})\alpha]r/2,$$

$$x_{B(2k-1)} = [(1-f)\cos(k-1)\alpha + (1+f)\cos(k-\tfrac{1}{2})\alpha]r/2,$$

$$y_{B(2k-1)} = [(1-f)\sin(k-1)\alpha + (1+f)\sin(k-\tfrac{1}{2})\alpha]r/2,$$

$$x_{A(2k)} = [(1+f)\cos k\alpha + (1-f)\cos(k-\tfrac{1}{2})\alpha]r/2,$$

$$y_{A(2k)} = [(1+f)\sin k\alpha + (1-f)\sin(k-\tfrac{1}{2})\alpha]r/2,$$

$$x_{B(2k)} = [(1-f)\cos k\alpha + (1+f)\cos(k-\tfrac{1}{2})\alpha]r/2, \text{ and}$$

$$y_{B(2k)} = [(1-f)\sin k\alpha + (1+f)\sin(k-\tfrac{1}{2})\alpha]r/2.$$

At this point, it will be easier to introduce some substitutions similar to what was done in Chapter 24 to simplify the algebraic manipulations. Let $s_1 = \sin k\alpha$, $c_1 = \cos$

Chapter 32. Bi/bigeneral tectal polyhedra II. Mathematical details. [Optional]

$k\alpha$, $s_2 = \sin \alpha/2$, and $c_2 = \cos \alpha/2$. Then

$$\sin \alpha = 2s_2c_2,$$
$$\cos \alpha = 2c_2^2 - 1,$$
$$\sin (k - \tfrac{1}{2})\alpha = s_1c_2 - c_1s_2,$$
$$\cos (k - \tfrac{1}{2})\alpha = c_1c_2 + s_1s_2,$$
$$\sin (k - 1)\alpha = s_1(2c_2^2 - 1) - 2c_1s_2c_2, \text{ and}$$
$$\cos (k - 1)\alpha = c_1(2c_2^2 - 1) + 2s_1s_2c_2.$$

With these substitutions, the coordinates become

$$x_{A(2k-1)} = (2c_1c_2^2 - c_1 + 2s_1s_2c_2 + f2c_1c_2^2 - fc_1 + 2fs_1s_2c_2 + c_1c_2 + s_1s_2 - fc_1c_2 - fs_1s_2)r/2,$$
$$y_{A(2k-1)} = (2s_1c_2^2 - s_1 - 2c_1s_2c_2 + 2fs_1c_2^2 - fs_1 - 2fc_1s_2c_2 + s_1c_2 - c_1s_2 - fs_1c_2 + fc_1s_2)r/2,$$
$$x_{B(2k-1)} = (2c_1c_2^2 - c_1 + 2s_1s_2c_2 - 2fc_1c_2^2 + fc_1 - 2fs_1s_2c_2 + c_1c_2 + s_1s_2 + fc_1c_2 + fs_1s_2)r/2,$$
$$y_{B(2k-1)} = (2s_1c_2^2 - s_1 - 2c_1s_2c_2 - 2fs_1c_2^2 + fs_1 + 2fc_1s_2c_2 + s_1c_2 - c_1s_2 + fs_1c_2 - fc_1s_2)r/2,$$
$$x_{A(2k)} = (c_1 + fc_1 + c_1c_2 + s_1s_2 - fc_1c_2 - fs_1s_2)r/2,$$
$$y_{A(2k)} = (s_1 + fs_1 + s_1c_2 - c_1s_2 - fs_1c_2 + fc_1s_2)r/2,$$
$$x_{B(2k)} = (c_1 - fc_1 + c_1c_2 + s_1s_2 + fc_1c_2 + fs_1s_2)r/2,$$

and

$$y_{B(2k)} = (s_1 - fs_1 + s_1c_2 - c_1s_2 + fs_1c_2 - fc_1s_2)r/2.$$

From these coordinates, the next step, which will not actually be demonstrated, is to determine the equality and parallelism of the appropriate edges, so that the fact that the polyhedron satisfies the definition in Chapter 31 of a quasiprism can be verified. This can be demonstrated, but to go through the actual algebraic manipulations would, as stated earlier, make this chapter so long that it has been foregone here.

Chapter 33. General notes on tectal polyhedra.

A number of tectal polyhedra have been defined and described in previous chapters. The most important symmetry-related properties are summarized in Table 18 below, in the same format as the previous Tables 7, 9, and 14.

Name of polyhedron	Symmetry group	n-Isohedral n	n-Isogonal n	n-Isotoxal n
prism	D_{nh}	2	1	2
antiprism	S_{2nv}	2	1	2
prismoid	C_{nv}	3	2	3
monocropped antiprism	C_{nv}	4	2	4
bicropped antiprism (quasiprism)	S_{2nv}	2	1	3
converged antiprism	S_{2nv}	4	2	3
truncated prism	D_{nh}	3	2	4
rectified prism	D_{nh}	3	2	2
truncated bipyramid	D_{nh}	3	3	3
cupola	C_{nv}	4	2	4
rotunda	C_{nv}	5	3	5
orthobicupola	D_{nh}	3	2	4
gyrobicupola	S_{2nv}	3	2	4
orthobirotunda	D_{nh}	4	3	5
gyrobirotunda	S_{2nv}	4	3	5
truncated pyramid	C_{nv}	4	3	5

Table 18: Primary symmetry properties of some tectal polyhedra.

Appendix A. Useful trigonometric formulas.

This appendix gives the derivation of some formulas which are used in a number of places (in the optional chapters only, however) in this book. Those readers who do not intend to read the optional chapters are advised to omit this and the following appendix. Three formulas will be assumed in these derivations:

$$\sin^2 \theta + \cos^2 \theta = 1,$$
$$\sin(\alpha + \beta) = \sin \alpha \cos \beta + \cos \alpha \sin \beta, \text{ and}$$
$$\cos(\alpha + \beta) = \cos \alpha \cos \beta - \sin \alpha \sin \beta.$$

Let $c = \cos \alpha$ and $s = \sin \alpha$. (This simplifies the algebraic manipulations.) Because of the first of the three equations above, one can always replace s^2 by $1 - c^2$ or c^2 by $1 - s^2$ whenever it makes the equation more convenient. (By these substitutions, all cosines of multiple angles can be expressed in terms of $\cos \alpha$ alone, while all sines of multiple angles can be expressed in terms of $\sin \alpha$ alone, except for *one* factor of $\cos \alpha$ appearing in certain multiples.) Then:

$$\cos 2\alpha = \cos(\alpha + \alpha) = c^2 - s^2$$
$$= c^2 - (1 - c^2)$$
$$= 2c^2 - 1.$$

$$\sin 2\alpha = \sin(\alpha + \alpha) = sc + cs$$
$$= 2sc.$$

$$\cos 3\alpha = \cos(\alpha + 2\alpha) = c(2c^2 - 1) - s(2sc)$$
$$= 2c^3 - c - 2s^2 c$$
$$= 2c^3 - c - 2(1 - c^2)c$$
$$= 2c^3 - c - 2c + 2c^3$$
$$= 4c^3 - 3c.$$

$$\sin 3\alpha = \sin(\alpha + 2\alpha) = s(2c^2 - 1) + c(2sc)$$
$$= 2sc^2 - s + 2sc^2$$
$$= 4sc^2 - s$$
$$= 4s(1 - s^2) - s$$
$$= 4s - 4s^3 - s$$
$$= 3s - 4s^3.$$

$$\cos 4\alpha = \cos(\alpha + 3\alpha) = c(4c^3 - 3c) - s(3s - 4s^3)$$
$$= 4c^4 - 3c^2 - 3s^2 + 4s^4$$

$$= 4c^4 - 3c^2 - 3(1-c^2) + 4(1-c^2)^2$$
$$= 4c^4 - 3c^2 - 3(1-c^2) + 4(1-2c^2+c^4)$$
$$= 4c^4 - 3c^2 - 3 + 3c^2 + 4 - 8c^2 + 4c^4$$
$$= 8c^4 - 8c^2 + 1$$

$$\sin 4\alpha = \sin(\alpha + 3\alpha) = s(4c^3 - 3c) + c(3s - 4s^3)$$
$$= 4sc^3 - 3sc + 3sc - 4s^3c$$
$$= 4sc^3 - 4s^3c$$
$$= 4sc(1 - s^2) - 4s^3c$$
$$= 4sc - 4s^3c - 4s^3c$$
$$= 4sc - 8s^3c$$
$$= 4sc(1 - 2s^2).$$

In other words, replacing s and c by $\cos \alpha$ and $\sin \alpha$, the multiple-angle formulas become:

$$\cos 2\alpha = 2\cos^2 \alpha - 1,$$

$$\sin 2\alpha = 2 \sin \alpha \cos \alpha,$$

$$\cos 3\alpha = 4\cos^3 \alpha - 3 \cos \alpha,$$

$$\sin 3\alpha = 3 \sin \alpha - 4 \sin^3 \alpha,$$

$$\cos 4\alpha = 8\cos^4 \alpha - 8\cos^2 \alpha + 1, \text{ and}$$

$$\sin 4\alpha = 4 \sin \alpha \cos \alpha (1 - 2 \sin^2 \alpha).$$

These formulas, for functions of multiple angles in terms of the functions of the single angle, will be used at a number of places in this book, in conjunction with the formulas for the Cartesian coordinates of a point in terms of the distance from the z-axis and angles of rotation about that axis:

$$x = r \cos \alpha,$$

$$y = r \sin \alpha.$$

Appendix B.   Concepts of vector algebra and geometry.

To a mathematician, the term *vector* means "any of the entities having certain properties in the construct known as a *vector space*," and "*vector space*" means a particular type of abstract algebraic structure which is studied in great detail in the branch of mathematics known as *linear algebra*. This book does not attempt to be a textbook of linear algebra, but merely, in this Appendix, presents some vector-related concepts that are useful in building the descriptions of the polyhedra which are the primary subject matter of the book. And for this purpose, this general sort of vector is not necessary.

For this book, the term *vector* will be taken as synonymous with geometric vector in 3-dimensional Euclidean space. A geometric vector, in turn, will be defined in terms of some other concepts that must be brought out first. Suppose one considers two points P_1 and P_2 (which could be two vertices of a polyhedron, but do not have to be; however, in this book they almost always will be either vertices of a polyhedron or other points that can be described in terms of the polyhedron geometry, such as the midpoint of an edge). For this purpose, it will be understood that the coordinates of P_1 are (x_1, y_1, z_1), and similarly for all other points in this discussion described as P_n, which will have coordinates (x_n, y_n, z_n). We can connect these points by a segment P_1P_2. If one specifies that the *start* of the segment is P_1 and the *end* is P_2, a direction is given to P_1P_2. The *directed segment* P_1P_2, thus, is different from (in fact, is considered the negative of) the directed segment P_2P_1. A geometric vector can be considered as *what is common to all* directed segments of the same length and direction. It should be noted that if there are two directed segments P_1P_2 and P_3P_4, with

$$x_2 - x_1 = x_4 - x_3,$$

$$y_2 - y_1 = y_4 - y_3, \text{ and}$$

$$z_2 - z_1 = z_4 - z_3,$$

then the directed segments P_1P_2 and P_3P_4 have the same length and direction, so that these coordinate differences $x_2 - x_1$, $y_2 - y_1$, and $z_2 - z_1$ completely characterize the geometric vector that has been defined by what is common to all directed segments of the same length and direction as P_1P_2. These coordinate differences are called the *components* of the vector **P_1P_2**. (Note that vectors are represented in this discussion by boldface letters.)

Thus, supposing the two points P_1 and P_2 to be given by P_1: (5, -6, 8) and P_2: (7, 0, 1). The directed segment from point P_1 to P_2 has an *x*-component of 2, a *y*-component of 6, and a *z*-component of -7; it can be symbolized as (2, 6, -7). Although it is impossible to distinguish directed segments or vectors written in terms of components from points written in terms of coordinates, it should always be clear whether a point, a directed segment, or a vector is meant from context.

Further, suppose there are two more points P_3 and P_4, given by P_3: (0, -2, 4) and P_4: (2, 4, -3). The directed segment from point P_3 to P_4 can also be seen to have components (2, 6, -7). Although the directed segments P_1P_2 and P_3P_4 are not the same segment (because they are in different parts of space), the *vectors* **P_1P_2** and **P_3P_4** *are* considered the same vector. The *components* are what matters, and they are *common to all* directed segments of the same length and direction, as described above. A vector may be specified by naming the endpoints of a directed segment which is an instance of the vector, but it must be understood that (boldface being used to imply a vector) the vectors **P_1P_2** and **P_3P_4** mean the same vector, and the difference in the points used to specify them is irrelevant. A vector may also be named by a boldface small letter such as **v**.

Vectors can be added or subtracted by adding or subtracting components. If two vectors **v** and **w** are given by **v**: (1, -1, 0) and **w**: (2, 2, 2), the sum **v** + **w** = (3, 1, 2). The difference, similarly, is **v** − **w** = (-1, -3, -2). A geometric interpretation of the sum is as follows:

1. Begin the first vector at any point, since the start point is irrelevant and only the relationship between the start and end points matters,
2. Begin the second vector at the end point of the first, and
3. The vector from the start point of the first to the end point of the second is the sum.

There are other ways, but this is the simplest to describe.

It is also possible to multiply a vector by an ordinary number. (Ordinary numbers are generally referred to as *scalars* when it is desirable to distinguish them from vectors in a discussion that deals with both.) It would, of course, be desirable to have a definition such that 2**v** = **v** + **v**, 3**v** = 2**v** + **v**, etc., and this is achieved by simply defining the product of a scalar and a vector as a new vector with each component obtained by multiplying the corresponding component of the original vector by that scalar. In other words, if the components of **v** are (x_v, y_v, z_v), the components of k**v** are (kx_v, ky_v, kz_v). This is the standard definition, used universally.

One thing that is always true, and will be used at some points in this book, is that if there are four points P_1, P_2, P_3, and P_4, such that the vector **P_3P_4** is a scalar multiple of the vector **P_1P_2**, the line segments P_1P_2 and P_3P_4 are parallel. This will be used, when it is convenient, to prove edges of a polyhedron to be parallel.

Since the components of a vector have been defined in terms of coordinate differences between points, so that, given points P_1: (x_1, y_1, z_1) and P_2: (x_2, y_2, z_2), the vector **P_1P_2** is given by $(x_2 - x_1, y_2 - y_1, z_2 - z_1)$, it is possible to work the other way, which means that, given a point P_1: (x_1, y_1, z_1), and knowing that the components of a vector **P_1P_2** are (x_{12}, y_{12}, z_{12}), we can determine the coordinates of P_2: $(x_1 + x_{12}, y_1 + y_{12}, z_1 + z_{12})$. Furthermore, if we wish to locate P_3 along the line P_1P_2 such that **P_1P_3** = k**P_1P_2**, we can simply

Appendix B. Concepts of vector algebra and geometry.

set the coordinates of P_3: $(x_1 + kx_{12}, y_1 + ky_{12}, z_1 + kz_{12})$. In particular, if P_3 is to be the midpoint of P_1P_2, so $k = \frac{1}{2}$,

$$x_3 = x_1 + x_{12}/2$$
$$= x_1 + \frac{1}{2}(x_2 - x_1)$$
$$= (x_2 + x_1)/2.$$

Similar formulas apply to the *y*- and *z*-coordinates. Another calculation we will want to make is the following: Suppose we have a line segment P_1P_2 and we wish to locate two points P_3 and P_4 between them such that $P_1P_3 = P_4P_2 = fP_1P_2$. For this case we first locate P_3 by

$$x_3 = x_1 + fx_{12}$$
$$= x_1 + f(x_2 - x_1)$$
$$= fx_2 + (1 - f)x_1,$$

with similar formulas applying to the *y*- and *z*-coordinates. Then P_4 is located by

$$x_4 = x_1 + (1 - f)x_{12}$$
$$= x_1 + (1 - f)(x_2 - x_1)$$
$$= fx_1 + (1 - f)x_2,$$

again with similar formulas applying to the *y*- and *z*-coordinates. This formula demonstrates a symmetry which will be useful.

While one might wish to be able to multiply two vectors together, there have been several types of multiplication that have been defined. The only one that will be extensively utilized in this book is the operation called the "dot product" (because a dot is the common symbol used for it) or the "scalar product" (because this product produces a *scalar, not* a vector, when two vectors are multiplied by this rule). If the components of two vectors **v** and **v**′ with components given by

$$\mathbf{v}: (v_1, v_2, v_3) \text{ and}$$
$$\mathbf{v}': (v'_1, v'_2, v'_3),$$

the dot product, or scalar product, is given by

$$\mathbf{v} \cdot \mathbf{v}' = v_1 v'_1 + v_2 v'_2 + v_3 v'_3.$$

It can be shown that the length of any vector **v** is given by $|\mathbf{v}| = \sqrt{(\mathbf{v} \cdot \mathbf{v})}$, which is one reason for using this product, and also that

$$\mathbf{v} \cdot \mathbf{v}' = |\mathbf{v}| |\mathbf{v}'| \cos \theta,$$

where θ is the angle between the vectors, a fact that enables us to calculate angles, another important reason for the use of this product.

On page 136 it was stated that, while the earlier part of this discussion was framed in terms of a vector being determined by its starting and ending points, the definition can be used in the reverse manner: given a vector and its starting point, the ending point is determined, and frequently the definition is used that way. For example, if one fixes a starting point P and a vector $\mathbf{v}$, all the points Q such that the vector $\mathbf{PQ}$ is equal to $c\mathbf{v}$, for all c, define a line, and given a starting point P and two vectors $\mathbf{v}$ and $\mathbf{v}'$, all the points Q such that the vector $\mathbf{PQ}$ is equal to $c\mathbf{v} + c'\mathbf{v}'$, for all c and c', define a plane. These properties will be used in this book when it is necessary to ensure that a point lies on a line or in a plane. Specifically, if four points P_1, P_2, P_3, and P_4 are to form a plane quadrilateral, it must be ascertained that $P_1P_3 = cP_1P_2 + c'P_1P_4$, for some c and c'.

It is also possible to ensure that four points P_1, P_2, P_3, and P_4 form a plane quadrilateral by showing that the line segments P_1P_2 and P_3P_4 are parallel, which can be demonstrated by showing that the vectors $\mathbf{P_1P_2}$ and $\mathbf{P_3P_4}$ obey an equation of the form

$$\mathbf{P_1P_2} = c\mathbf{P_3P_4}$$

for some scalar constant c. And one specific case of this will be important enough that it will be demonstrated at this point, to be referred to elsewhere in the book as the *two-plane parallel rule*. For this, it will be assumed that there are four points P_1, P_2, P_3, and P_4, whose coordinates are given by

$$P_1: (r_1 \cos \theta, r_1 \sin \theta, z_1),$$

$$P_2: (r_1 \cos \phi, r_1 \sin \phi, z_1),$$

$$P_3: (r_2 \cos [\theta + \beta], r_2 \sin [\theta + \beta], z_2), \text{ and}$$

$$P_4: (r_2 \cos [\phi - \beta], r_2 \sin [\phi - \beta], z_2).$$

Geometrically, this means that P_1 and P_2 are in the same plane parallel to the xy-plane, and so are P_3 and P_4 (though the plane containing P_1 and P_2 is different from the plane containing P_3 and P_4, if $z_2 \neq z_1$). In addition, P_1 and P_2 are on a circle in their common plane parallel to the xy-plane, centered on the point where the z-axis pierces that plane, and so are P_3 and P_4 (though the radii of these planes are different, if $r_2 \neq r_1$). And finally, if one draws a plane through the z-axis, cutting *either* the plane containing P_1 and P_2 or the plane containing P_3 and P_4 halfway between the two points in question, it will also cut the other of those two planes halfway between the two points which defined that plane.

Appendix B. Concepts of vector algebra and geometry.

The *two-plane parallel rule* states that when these conditions are met, the line segments P_1P_2 and P_3P_4 (or for that matter, the lines containing those segments) are parallel.

To demonstrate this, first compute the vectors P_1P_2 and P_3P_4. It will then be shown that they obey an equation of the form $P_1P_2 = cP_3P_4$.

$$P_1P_2 = (r_1[\cos\phi - \cos\theta], r_1[\sin\phi - \sin\theta], 0), \text{ and}$$

$$P_3P_4 = (r_2\{\cos[\phi - \beta] - \cos[\theta + \beta]\}, r_2\{\sin[\phi - \beta] - \sin[\theta + \beta]\}, 0).$$

To show these to be parallel, it is necessary to show that

$$\cos[\phi - \beta] - \cos[\theta + \beta] = k(\cos\phi - \cos\theta) \text{ and}$$

$$\sin[\phi - \beta] - \sin[\theta + \beta] = k(\sin\phi - \sin\theta),$$

for a single value of k. Now, using the addition and subtraction formulas for the trigonometric functions, as given at the beginning of Appendix A, one can write

$$\cos[\phi - \beta] - \cos[\theta + \beta] = \cos\phi\cos\beta + \sin\phi\sin\beta - \cos\theta\cos\beta + \sin\theta\sin\beta$$

$$= \cos\phi\cos\beta - \cos\theta\cos\beta + \sin\phi\sin\beta + \sin\theta\sin\beta$$

$$= (\cos\phi - \cos\theta)\cos\beta + (\sin\phi + \sin\theta)\sin\beta,$$

$$\sin[\phi - \beta] - \sin[\theta + \beta] = \sin\phi\cos\beta - \cos\phi\sin\beta - \sin\theta\cos\beta - \cos\theta\sin\beta$$

$$= \sin\phi\cos\beta - \sin\theta\cos\beta - \cos\phi\sin\beta - \cos\theta\sin\beta$$

$$= (\sin\phi - \sin\theta)\cos\beta - (\cos\phi + \cos\theta)\sin\beta.$$

It can be seen that what is necessary to show is that

$$(\cos\phi - \cos\theta)\cos\beta + (\sin\phi + \sin\theta)\sin\beta = k(\cos\phi - \cos\theta) \text{ and}$$

$$(\sin\phi - \sin\theta)\cos\beta - (\cos\phi + \cos\theta)\sin\beta = k(\sin\phi - \sin\theta),$$

for the same value of k. These two equations can more simply be written as

$$(\sin\phi + \sin\theta)\sin\beta = (k - \cos\beta)(\cos\phi - \cos\theta) \text{ and}$$

$$-(\cos\phi + \cos\theta)\sin\beta = (k - \cos\beta)(\sin\phi - \sin\theta),$$

which, eliminating the common factors $\sin\beta$ and $(k - \cos\beta)$, requires that

$$(\sin\phi + \sin\theta)(\sin\phi - \sin\theta) = -(\cos\phi + \cos\theta)(\cos\phi - \cos\theta).$$

But the left-hand side of this equation is simply $\sin^2\phi - \sin^2\theta$, and similarly the right-hand sign is $-(\cos^2\phi - \cos^2\theta) = -[(1 - \sin^2\phi) - (1 - \sin^2\theta)]$, which simplifies to the same expression, $\sin^2 \phi - \sin^2 \theta$. So the necessary formula is demonstrated. More specifically, it has been shown that, if

$$\lambda = (\sin \phi + \sin \theta)(\sin \phi - \sin \theta) = -(\cos \phi + \cos \theta)(\cos \phi - \cos \theta),$$

both of which have been shown to be equal, one can write

$$\sin \phi + \sin \theta = \lambda/(\sin \phi - \sin \theta) \text{ and}$$

$$\cos \phi + \cos \theta = -\lambda/(\cos \phi - \cos \theta),$$

and then, combining this with either of the two equations

$$(\sin \phi + \sin \theta)\sin \beta = (k - \cos \beta)(\cos \phi - \cos \theta) \text{ and}$$

$$-(\cos \phi + \cos \theta)\sin \beta = (k - \cos \beta)(\sin \phi - \sin \theta),$$

one can derive

$$\lambda \sin \beta/(\sin \phi - \sin \theta) = (k - \cos \beta)(\cos \phi - \cos \theta).$$

This will give

$$(k - \cos \beta)(\cos \phi - \cos \theta)(\sin \phi - \sin \theta) = \lambda \sin \beta$$

or

$$k = \cos \beta + \lambda \sin \beta/(\cos \phi - \cos \theta)(\sin \phi - \sin \theta).$$

Substituting the definition of λ above, one has

$$k = \cos \beta + (\sin \phi + \sin \theta)\sin \beta/(\cos \phi - \cos \theta)$$

$$= \cos \beta - (\cos \phi + \cos \theta)\sin \beta/(\sin \phi - \sin \theta),$$

both expressions having been shown to be equivalent. So in fact, it has been shown, not only that $P_1P_2 \parallel P_3P_4$, but also that $P_3P_4 = kP_1P_2$, with k given by either of the two expressions just shown. This is the *two-plane parallel rule* which will be used in this book.

Appendix C.   Coordinates of the Platonic solids 1. The tetrahedron.

Various modification procedures have been described in the chapters of this book entitled "Modification procedures involving polyhedra." In order to make it easy to generate the coordinates of the vertices of different polyhedra that can be obtained by such modification procedures, this and the following four Appendices are provided.

Each of the five Appendices will have lists of coordinates. In most (but not all) of the lists, ± signs are found; this means that all possible choices are to be made, each giving a different point. So, if (0, ±1, ±1) appears, this means that four possibilities are represented: Each of the two ± signs may be taken as positive or negative, independently.

Sometimes, when the vertices are listed in a way that makes it convenient, the system of labeling the points according to vertex rotation plane (described on p. 69, Chapter 18) is used. In other cases, they will simply be listed as P_1, P_2, etc. (When, say, four different points are represented by a set of coordinates because of the presence of ± signs as in the previous paragraph, a designation such as P_9-P_{12} will appear.)

It will be convenient to give more than one set of coordinates, because for each polyhedron there may be more than one orientation that makes sense to use.

Coordinates will be given in a form that makes the numbers convenient to recognize. This will mean that not all the polyhedra are the same size, but for each coordinate set, the length of the edge is provided, and all coordinates may be scaled with this information.

In this Appendix, some sets of coordinates for the vertices of the tetrahedron are given. These can be used as starting points, in conjunction with the modification procedures described in the book, to generate other polyhedron coordinates.

Set 1.

While one most frequently thinks of a tetrahedron as a triangular pyramid, it is actually easier to generate the coordinates of the vertices if one concentrates on the fourfold alternating axis and thinks of it as a digonal antiprism. There are two natural ways to assign coordinates to the vertices on this basis. In the first one, it is usual to start with the eight vertices of a cube, to be given in Appendix F. If one selects four of these properly, the vertices of a tetrahedron are obtained. Either the four with an odd number of *positive* signs or the four with an odd number of *negative* signs can be chosen; this set will give the case of an odd number of positive signs, simply because that way the "all-three-positive" vertex at (+1, +1, +1) is included. This gives:

P_1: (+1, +1, +1)

P_2: (−1, −1, +1)

P_3: (−1, +1, −1)

P_4: (+1, −1, −1)

It can be seen that one can choose any plane parallel to a coordinate plane passing through one of these vertices, and it will include one other vertex, both oppositely directed from the local origin (*i. e.*, the point in that plane where the coordinate axis perpendicular to the plane intersects it).

Any two vertices define an edge. Each edge has a length of $2\sqrt{2}$. Any three vertices define a face. One can calculate the location of the centroid of a face by averaging the coordinates of the vertices of that face:

C_{123}: $((+1 - 1 + 1)/3, (+1 - 1 - 1)/3, (+1 + 1 - 1)/3)$

$= (-1/3, +1/3, +1/3)$

C_{124}: $((+1 - 1 - 1)/3, (+1 - 1 + 1)/3, (+1 + 1 - 1)/3)$

$= (+1/3, -1/3, +1/3)$

C_{134}: $((+1 + 1 - 1)/3, (+1 - 1 + 1)/3, (+1 - 1 - 1)/3)$

$= (+1/3, +1/3, -1/3)$

C_{234}: $((-1 + 1 - 1)/3, (-1 - 1 + 1)/3, (+1 - 1 - 1)/3)$

$= (-1/3, -1/3, -1/3)$

These, of course, except for an extra factor of $1/3$, are the four cubic vertices that were *not* chosen for the tetrahedron vertices. This illustrates that the tetrahedron is its own dual, but the dual tetrahedron is oppositely oriented from the original. By this procedure, thus, the dual of the dual is the original tetrahedron reduced in size by a factor of 9, so one might prefer, instead of using the centroids, to *scale* the dualization by using the centroids as a starting basis, and multiplying by 3.

Set 2.

In this set, a fourfold alternating axis will again be chosen as the z-axis, as in the previous case, but with the coordinate planes oriented so the xz- and yz-planes each pass through two vertices. This means that the z-scale has to be chosen carefully to keep all the edges equal:

P_1: (+1, 0, +½√2)

P_2: (−1, 0, +½√2)

P_3: (0, +1, −½√2)

Appendix C. Coordinates of the Platonic solids I. The tetrahedron.

P_4: $(0, -1, -\frac{1}{2}\sqrt{2})$

Each edge of this tetrahedron is equal to 2.

Appendix D. Coordinates of the Platonic solids II. The octahedron.

In this Appendix, some sets of coordinates for the vertices of the octahedron are given. These can be used as starting points, in conjunction with the modification procedures described in the book, to generate other polyhedron coordinates.

Set 1.

The simplest set of coordinates for the octahedron is obtained by orienting it so that the three fourfold axes all lie on the coordinate axes. The center will then be at the origin, and the coordinates will be given by:

P_1-P_2: (± 1, 0, 0)

P_3-P_4: (0, ± 1, 0)

P_5-P_6: (0, 0, ± 1)

Any edge can be generated by taking one vertex from one line of the three above, and one from either of the remaining lines. Since the positive or the negative sign may be chosen in each case, and the two lines can be chosen out of these three in any of three ways, this makes twelve edges. Each edge has a length of $\sqrt{2}$.

The threefold axes go through the origin and opposite pairs of points selected from the eight at (± 1, ± 1, ± 1). (Each pair is chosen by picking one of the eight possible combinations of $+$ and $-$ signs for one point, and reversing all the signs for the other. While there are eight ways to do this, in each case, the second point, gotten by reversing the signs of the coordinates of the first, is obtained as the first point in another choice, so the "eight" counts each axis twice. Thus, four threefold axes are found.)

Each face is obtained by taking one vertex from each of the three lines in the list. as there are 2 ways to do so for each vertex, all eight faces are generated.

The twofold axes are harder to spot. However, it can be seen that any point whose coordinates contain two ½'s and a zero (with any signs on the ½'s) is the midpoint of an edge. Two such points are opposite if the signs of one are reversed from the other. Note that which coordinate is zero can be chosen in three ways, and the signs of the two ½'s can be determined in four ways for each choice, but in each case one of the four will be the exact reversal of the signs of another of the four, so that once the choice of the coordinate to be made zero is made, there are two ways of determining an axis; therefore six twofold axes are all found by this procedure.

Set 2.

Another way of deriving the coordinates of the octahedron is to treat it as a triangular antiprism and use the derivation of Chapter 6. Since the value of α is equal to $2\pi/3$, one can use the known values of the trigonometric functions of multiples of $\alpha/2 = \pi/3$:

Appendix D. Coordinates of the Platonic solids II. The octahedron.

$$\sin \pi/3 = \sin 2\pi/3 = \tfrac{1}{2}\sqrt{3}$$
$$\sin 4\pi/3 = \sin 5\pi/3 = -\tfrac{1}{2}\sqrt{3}$$
$$\cos \pi/3 = \cos 5\pi/3 = \tfrac{1}{2}$$
$$\cos 2\pi/3 = \cos 4\pi/3 = -\tfrac{1}{2}$$

giving the coordinates:

A_1: $(r, 0, z_0)$

A_2: $(-\tfrac{1}{2}r, \tfrac{1}{2}r\sqrt{3}, z_0)$

A_3: $(-\tfrac{1}{2}r, -\tfrac{1}{2}r\sqrt{3}, z_0)$

B_1: $(\tfrac{1}{2}r, \tfrac{1}{2}r\sqrt{3}, -z_0)$

B_2: $(-r, 0, -z_0)$

B_3: $(\tfrac{1}{2}r, -\tfrac{1}{2}r\sqrt{3}, -z_0)$,

where if the edge of the octahedron is 1,

$$r = 1/\sqrt{[2(1 - \cos \alpha)]} = (\sqrt{3})/3, \text{ and}$$
$$z_0 = h/2 = \tfrac{1}{2}\sqrt{[(\cos \alpha/2 - \cos \alpha)/(1 - \cos \alpha)]} = (\sqrt{6})/6.$$

These results correspond, in fact, to the *n*=3 line of Table 5.

Appendix E. Coordinates of the Platonic solids III. The icosahedron.

In this Appendix, some sets of coordinates for the vertices of the icosahedron are given. These can be used as starting points, in conjunction with the modification procedures described in the book, to generate other polyhedron coordinates. The various ways one can consider the icosahedron determines how one will orient it, as was seen with the other Platonic solids in the previous two Appendices. For example, one can consider it as a biaugmented pentagonal antiprism. This will focus on one of the six fivefold axes, and one will end up with a construction with the z-axis along one of those fivefold axes. (Parenthetically, this structure can also be considered a gyroelongated pentagonal bipyramid, because both definitions envision a pentagonal antiprism fused to pyramids at base and roof). It is also possible to orient the icosahedron with the z-axis along one of the twofold or threefold axes. The orientation along the twofold axis produces a particularly interesting set of coordinates.

Set 1.

The easiest way to generate the coordinates of the vertices of the icosahedron is to treat it, as described in the previous paragraph, as a biaugmented pentagonal antiprism or gyroelongated pentagonal bipyramid. So one starts off by determining the vertices of a pentagonal antiprism, which are best designated B_1-B_5 and C_1-C_5 using vertex rotation plane based labeling. The apex is designated A and the antiapex D. The z-axis is chosen to run along the line AD, with the origin halfway between A and D. With this convention, the coordinates can be assigned as:

A: $(0, 0, h_1)$,

B_1: $(r, 0, h_2)$,

B_2: $(r \cos 2\pi/5, r \sin 2\pi/5, h_2)$,

B_3: $(r \cos 4\pi/5, r \sin 4\pi/5, h_2)$,

B_4: $(r \cos 4\pi/5, -r \sin 4\pi/5, h_2)$,

B_5: $(r \cos 2\pi/5, -r \sin 2\pi/5, h_2)$,

C_1: $(r \cos \pi/5, r \sin \pi/5, -h_2)$,

C_2: $(r \cos 3\pi/5, r \sin 3\pi/5, -h_2)$,

C_3: $(-r, 0, -h_2)$,

C_4: $(r \cos 3\pi/5, -r \sin 3\pi/5, -h_2)$,

C_5: $(r \cos \pi/5, -r \sin \pi/5, -h_2)$,

D: $(0, 0, -h_1)$.

Appendix E. Coordinates of the Platonic solids III. The icosahedron.

Here it is necessary to determine h_1 and h_2, which is done by making sure that the distances AB_1, B_1B_2, and B_1C_1 are all equal. But first, it should be noted that

$\sin \pi/5 = \frac{1}{4}[\sqrt{(10 - 2\sqrt{5})}]$,

$\cos \pi/5 = (1 + \sqrt{5})/4$,

$\sin 2\pi/5 = \frac{1}{4}[\sqrt{(10 + 2\sqrt{5})}]$, and

$\cos 2\pi/5 = (\sqrt{5} - 1)/4$.

(While it is also true that $\sin 3\pi/5 = \sin 2\pi/5$, $\cos 3\pi/5 = -\cos 2\pi/5$, $\sin 4\pi/5 = \sin \pi/5$, and $\cos 4\pi/5 = -\cos \pi/5$, those equivalences are unnecessary to calculate the distances AB_1, B_1B_2, and B_1C_1.)

Using these, it is possible to determine:

$$AB_1 = \sqrt{[r^2 + (h_1 - h_2)^2]},$$

$$B_1B_2 = r\sqrt{[2(1 - \cos 2\pi/5)]}$$

$$= r\sqrt{\{[2 + (1 - \sqrt{5})/2]\}}$$

$$= r\sqrt{[(5 - \sqrt{5})/2]}, \text{ and}$$

$$B_1C_1 = \sqrt{[2r^2(1 - \cos \pi/5) + 4h_2^2]}$$

$$= \sqrt{\{2r^2[1 - (1 + \sqrt{5})/4] + 4h_2^2\}}$$

$$= \sqrt{[(3 - \sqrt{5})r^2/2 + 4h_2^2]}.$$

The next step is then to set $B_1B_2 = B_1C_1$, or more easily $B_1B_2^2 = B_1C_1^2$, which requires

$$(5 - \sqrt{5})r^2/2 = (3 - \sqrt{5})r^2/2 + 4h_2^2, \text{ or}$$

$$r^2 = 4h_2^2; \text{ that is,}$$

$$h_2 = r/2.$$

The final step is then to set $AB_1 = B_1B_2$, or more easily $AB_1^2 = B_1B_2^2$, which requires

$$r^2 + (h_1 - h_2)^2 = (5 - \sqrt{5})r^2/2,$$

$$r^2 + (h_1 - r/2)^2 = (5 - \sqrt{5})r^2/2,$$

$$r^2 + (h_1^2 - rh_1 + r^2/4) = (5 - \sqrt{5})r^2/2,$$

$$5r^2/4 + h_1^2 - rh_1 = (5 - \sqrt{5})r^2/2,$$

$$4h_1^2 - 4rh_1 - (5 - 2\sqrt{5})r^2 = 0,$$

$$h_1 = \tfrac{1}{2}[1 + \sqrt{(6 - 2\sqrt{5})}]r.$$

It should be noted that this procedure fixes the length of the edge of the icosahedron at $r\sqrt{[(5 - \sqrt{5})/2]}$. Substituting these values into the expressions above for the coordinates, and letting $r = 1$ (as any constant multiple of all these coordinates will give a valid set of coordinates), the coordinates obtained are:

A: $(0, 0, \tfrac{1}{2}\{1 + \sqrt{[6 - 2\sqrt{5}]}\})$,

B_1: $(1, 0, \tfrac{1}{2})$,

B_2: $([\sqrt{5} - 1]/4, \tfrac{1}{4}\{\sqrt{[10 + 2\sqrt{5}]}\}, \tfrac{1}{2})$,

B_3: $(-[1 + \sqrt{5}]/4, \tfrac{1}{4}\{\sqrt{[10 - 2\sqrt{5}]}\}, \tfrac{1}{2})$,

B_4: $(-[1 + \sqrt{5}]/4, -\tfrac{1}{4}\{\sqrt{[10 - 2\sqrt{5}]}\}, \tfrac{1}{2})$,

B_5: $([\sqrt{5} - 1]/4, -\tfrac{1}{4}\{\sqrt{[10 + 2\sqrt{5}]}\}, \tfrac{1}{2})$,

C_1: $([1 + \sqrt{5}]/4, \tfrac{1}{4}\{\sqrt{[10 - 2\sqrt{5}]}\}, -\tfrac{1}{2})$,

C_2: $(-[\sqrt{5} - 1], \tfrac{1}{4}\{\sqrt{[10 + 2\sqrt{5}]}\}, -\tfrac{1}{2})$,

C_3: $(-1, 0, -\tfrac{1}{2})$,

C_4: $(-[\sqrt{5} - 1], -\tfrac{1}{4}\{\sqrt{[10 + 2\sqrt{5}]}\}, -\tfrac{1}{2})$,

C_5: $([1 + \sqrt{5}]/4, -\tfrac{1}{4}\{\sqrt{[10 - 2\sqrt{5}]}\}, -\tfrac{1}{2})$, and

D: $(0, 0, -\tfrac{1}{2}\{1 + \sqrt{[6 - 2\sqrt{5}]}\})$.

Numerically, these coordinates are given by:

A: $(0.000000, 0.000000, 1.118034)$,

B_1: $(1.000000, 0.000000, 0.500000)$,

B_2: $(0.309017, 0.951057, 0.500000)$,

B_3: $(-0.809017, 0.587785, 0.500000)$,

B_4: $(-0.809017, -0.587785, 0.500000)$,

B_5: $(0.309017, -0.951057, 0.500000)$,

C_1: $(0.809017, 0.587785, -0.500000)$,

C_2: $(-0.309017, 0.951057, -0.500000)$,

C_3: $(-1.000000, 0.000000, -0.500000)$,

C_4: $(-0.309017, -0.951057, -0.500000)$,

Appendix E. Coordinates of the Platonic solids III. The icosahedron.

C_5: (0.809017, -0.587785, -0.500000), and

D: (0.000000, 0.000000, -1.118034).

Set 2.

An interesting set of coordinates, as was earlier stated, is obtained when the *z*-axis is oriented along a twofold axis of the icosahedron, the line joining the midpoints of two opposite edges of the icosahedron. It turns out that then the *x*- and *y*-axes also lie along twofold axes of the icosahedron, and the vertices then lie at the corners of three congruent rectangles, one each in the *xy*-, *xz*-, and *yz*-planes. (This will not be shown at this point, but assumed so, and one will see, when the twelve sets of coordinates of the vertices are computed, that they do form an icosahedron.)

Thus, the vertices can be initially set with coordinates as follows, with the rectangles being $P_1P_2P_3P_4$, $P_5P_6P_7P_8$, and $P_9P_{10}P_{11}P_{12}$:

P_1-P_4: ($\pm e/2$, $\pm m$, 0),

P_5-P_8: ($\pm m$, 0, $\pm e/2$), and

P_9-P_{12}: (0, $\pm e/2$, $\pm m$),

where *e* is the length of the edge of the icosahedron, and *m* is yet to be determined. (For ease in referring to individual points, the signs will be chosen so that the *first* point, namely P_1, P_5, and P_9, of each set of four has the signs both positive, the second has + on the *e*/2 and − on the *m*, the third has both negative, and the fourth has − on the *e*/2 and + on the *m*. Then automatically the sides P_1P_2, P_3P_4, P_5P_6, P_7P_8, P_9P_{10}, and $P_{11}P_{12}$ of the rectangles are equal to *e*, and these will form six of the thirty edges of the icosahedron. (The remaining six rectangle sides do not form edges of the final icosahedron.)

By calculation, the distances P_1P_5, P_1P_8, P_2P_5, and P_2P_8 all can be found to be equal to

$$\sqrt{[(m - e/2)^2 + m^2 + e^2/4]}$$
$$= \sqrt{[(m^2 - em + e^2/4) + m^2 + e^2/4]}$$
$$= \sqrt{(2m^2 - em + e^2/2)}.$$

If this is made equal to *e*, it requires

$$2m^2 - em + e^2/2 = e^2,$$
$$2m^2 - em - e^2/2 = 0,$$
$$4m^2 - 2em - e^2 = 0,$$

$$m = (\sqrt{5} + 1)e/4.$$

It might be noted that in fact $m = e\tau/2$, where τ is the golden ratio $= (\sqrt{5} + 1)/2 \approx 1.61803$ (see my book, *The Fibonacci Sequence and Beyond*, for my reasons for not using the symbol ϕ, which is often used), the positive number such that $\tau^2 = \tau + 1$, so each of the three rectangles $P_1P_2P_3P_4$, $P_5P_6P_7P_8$, and $P_9P_{10}P_{11}P_{12}$ has sides e and $2m=e\tau$, and is thus a golden rectangle.

With this value of m, it can in fact be seen that not just P_1P_5, P_1P_8, P_2P_5, and P_2P_8 are equal to e, but all the twenty-four distances connecting a vertex of a rectangle to the nearest vertices of the other two rectangles:

P_1P_5, P_1P_8, P_1P_9, P_1P_{10},

P_2P_5, P_2P_8, P_2P_{11}, P_2P_{12},

P_3P_6, P_3P_7, P_3P_{11}, P_3P_{12},

P_4P_6, P_4P_7, P_4P_9, P_4P_{10},

P_5P_9, P_5P_{12},

P_6P_9, P_6P_{12},

P_7P_{10}, P_7P_{11},

P_8P_{10}, and P_8P_{11}.

Together with the shorter sides P_1P_2, P_3P_4, P_5P_6, P_7P_8, P_9P_{10}, and $P_{11}P_{12}$ of the three rectangles, which were shown earlier top be equal to e, there are thirty distances all equal to the same value, forming twenty equilateral triangles $P_1P_2P_5$, $P_1P_5P_8$, etc., so that these points are, in fact, the vertices of a regular icosahedron.

Appendix F. Coordinates of the Platonic solids IV. The cube.

In this Appendix, some sets of coordinates for the vertices of the cube are given. These can be used as starting points, in conjunction with the modification procedures described in the book, to generate other polyhedron coordinates.

Set 1.

This set is the simplest one for the cube.

P_1-P_8: (± 1, ± 1, ± 1)

Each edge is bounded by two vertices that are specified by fixing two of the three $\pm$ signs (either as + or as −) in the above and allowing the remaining sign to take all possible values. It can be seen that each edge is equal to 2.

Each face has its four vertices determined by fixing *one* of the three $\pm$ signs in the above and allowing the other *two* signs to take all possible values.

The fourfold axes run along the three coordinate axes, as in the case of the regular octahedron which is located as described under "Set 1" in Appendix D, which will be termed the *Set 1 octahedron*. This orientation places the cube in a position that emphasizes the duality with that octahedron. For the centroid of the face whose vertices have coordinates (± 1, ± 1, +1), for example, is at (0, 0, +1), which is clearly one of the vertices of the Set 1 octahedron. Taking each face of this cube in turn, it can be seen that its centroid is a vertex of that octahedron.

Since this cube has been oriented the same way as the Set 1 octahedron, the other rotational axes can be located in the same way as for that octahedron: the threefold axes pass through opposite sets of vertices, obtained by picking one of the eight possible choices of signs for the first vertex and reversing each sign to get the opposite vertex, while the twofold axes pass through the edge midpoints, which actually are twice as far from the origin as the edge midpoints of the octahedron (thus they include all points with two 1's and a zero, choosing any signs for the 1's). Since any of the three coordinates can be chosen to be the zero, and the two 1's can be assigned signs in four different ways, twelve edge midpoints are found, which form six pairs of opposite points, and these determine six twofold axes.

Consider one face of the Set 1 octahedron, say the one bounded by (1, 0, 0), (0, 1, 0), and (0, 0, 1) (Any other choice of signs will work as well; the all-positive case was chosen for greatest simplicity.) The centroid of this face is at ($1/3$, $1/3$, $1/3$), which is only $1/3$ as far from the origin as in the cube described here, whose dual was seen to be the Set 1 octahedron. Thus in fact, the dual of the dual of that octahedron is only $1/3$ of its size, which suggests that one might prefer to consider each dualization scaled by a factor of $\sqrt{3}$: instead of using the centroid of each face as a vertex of the dual, use the point $\sqrt{3}$ times as far from the origin. Note that in Appendix C, a similar consideration for the *tetrahedron* suggested that instead of using the centroid of each face as a vertex of the dual, one might prefer to use the point 3 times as far from the origin. For the cube/octahedron pair, thus, a much smaller degree of scaling is necessary.

Set 2.

For this set, the cube will be considered as a triangular antibipyramid, as noted on p. 81 in Chapter 21, oriented so that one of the threefold axes is the z-axis. Using the vertex rotation plane based labeling of Chapter 18, and a logical orientation of the x-axis, it can be seen that the vertices are at:

A: $(0, 0, h_1)$,

B_1: $(r, 0, h_2)$,

B_2: $(-r/2, r\sqrt{3}/2, h_2)$,

B_3: $(-r/2, -r\sqrt{3}/2, h_2)$,

C_1: $(r/2, r\sqrt{3}/2, -h_2)$,

C_2: $(-r, 0, -h_2)$,

C_3: $(r/2, -r\sqrt{3}/2, -h_2)$,

D: $(0, 0, -h_1)$.

It is necessary to fix the ratios of h_1, h_2, and r so all the angles are 90°. First of all, the length of AB_1 is given by $\sqrt{[r^2 + (h_1 - h_2)^2]}$, as is the length of AB_2. Using the vector dot product described in Appendix B,

$$AB_1 \cdot AB_2 = -r^2/2 + (h_2 - h_1)^2,$$

or, since the square of any quantity is the square of its negative, and $h_1 > h_2$, we might prefer to write

$$AB_1 \cdot AB_2 = -r^2/2 + (h_1 - h_2)^2.$$

But $AB_1 \cdot AB_2 = |AB_1||AB_2| \cos(B_1AB_2)$, and to make $B_1AB_2 = 90°$, we must have $\cos(B_1AB_2) = 0$. So

$$-r^2/2 + (h_1 - h_2)^2 = 0,$$

$$(h_1 - h_2)^2 = r^2/2,$$

$$r^2 = 2(h_1 - h_2)^2,$$

$$r = (h_1 - h_2)\sqrt{2}.$$

The length of B_1C_1 is given by:

Appendix F. Coordinates of the Platonic solids IV. The cube.

$$\sqrt{[(r/2)^2 + (r\sqrt{3}/2)^2 + (2h_2)^2]}$$
$$= \sqrt{[(r^2/4) + (3r^2/4) + 4h_2^2]}$$
$$= \sqrt{(r^2 + 4h_2^2)}.$$

But the length of AB_1 is given by $\sqrt{[r^2 + (h_1 - h_2)^2]}$, and for those to be equal,

$$4h_2^2 = (h_1 - h_2)^2, \text{ or}$$
$$2h_2 = (h_1 - h_2), \text{ implying that}$$
$$h_1 = 3h_2.$$

Putting all these results together, one has, using h instead of h_2 since only one remains after simplifying:

A: $(0, 0, 3h)$,

B_1: $(2h\sqrt{2}, 0, h)$,

B_2: $(-h\sqrt{2}, h\sqrt{6}, h)$,

B_3: $(-h\sqrt{2}, -h\sqrt{6}, h)$,

C_1: $(h\sqrt{2}, h\sqrt{6}, -h)$,

C_2: $(-2h\sqrt{2}, 0, -h)$,

C_3: $(h\sqrt{2}, -h\sqrt{6}, -h)$,

D: $(0, 0, -3h)$.

Appendix G. Coordinates of the Platonic solids V. The dodecahedron.

In this Appendix, some sets of coordinates for the vertices of the dodecahedron are given. These can be used as starting points, in conjunction with the modification procedures described in the book, to generate other polyhedron coordinates.

Set 1.

This set is the most frequently encountered in the literature, although it does not show clearly the edges of the figure. The coordinate axes are along twofold axes of rotation of the dodecahedron; the fivefold axes are not easy to spot, but *one* of the threefold axes is through the origin, connecting the points (-1, -1, -1) and (1, 1, 1).

P_1-P_8: (± 1, ± 1, ± 1)

P_9-P_{12}: (0, $\pm\tau$, $\pm[\tau - 1]$)

P_{13}-P_{16}: ($\pm\tau$, $\pm[\tau - 1]$, 0)

P_{17}-P_{20}: ($\pm[\tau - 1]$, 0, $\pm\tau$)

where, as in Appendix E, τ represents the golden ratio. It should be noted that the eight points P_1-P_8 lie at the vertices of a cube (see Appendix F), while each of the remaining sets of four points P_9-P_{12}, P_{13}-P_{16}, and P_{17}-P_{20} lie at the vertices of a rectangle. (Since $\tau - 1 = 1/\tau$, these rectangles have the proportion of $1{:}\tau^2$, which can also written as $1{:}\tau + 1$.)

The dodecahedron specified by these twenty vertex coordinates has an edge length equal to $\sqrt{5} - 1 = 2(\tau - 1) \approx 1.23607$. In each of the three rectangles specified by P_9-P_{12}, P_{13}-P_{16}, and P_{17}-P_{20}, the two shorter sides of the rectangle are edges of the dodecahedron, accounting for six of the thirty edges. In addition, each of the eight vertices P_1-P_8 is connected by an edge to whichever of the twelve vertices P_9-P_{20} has the same signs for its nonzero coordinates. Thus (+1, +1, +1) is connected to the three points (0, +τ, +[$\tau - 1$]), (+τ, +[$\tau - 1$], 0), and (+[$\tau - 1$], 0, +τ). This accounts for the remaining twenty-four of the edges.

Set 2.

Although the coordinates listed in Set 1 are the most commonly found, perhaps because they make clear that eight vertices of the dodecahedron lie on the vertices of a cube (actually, the twenty vertices can be grouped into five sets of eight which define cubes, each one in two different sets, but this is less obvious), it is really difficult to use these coordinates to draw, say, a picture of the dodecahedron. The best set of coordinates for this purpose takes the dodecahedron as a bifrustum of a pentagonal antibipyramid, and utilizes this to generate the coordinates. With the parameters chosen in such a way as to make all the pentagons regular and equal, the coordinates come out as:

A_1-A_5: (cos [$n - 1$]α, sin [$n - 1$]α, $1 + \sin \alpha/4$) where n = 1 to 5 and $\alpha = 72° = 2\pi/5$ radians

Appendix G. Coordinates of the Platonic solids V. The dodecahedron.

B_1-B_5: $(2 \cos[n-1]\alpha \cos \alpha/2,\ 2 \sin[n-1]\alpha \cos \alpha/2,\ \sin \alpha/4)$

C_1-C_5: $(2 \cos[n-\tfrac{1}{2}]\alpha \cos \alpha/2,\ 2 \sin[n-\tfrac{1}{2}]\alpha \cos \alpha/2,\ -\sin \alpha/4)$

D_1-D_5: $(\cos[n-\tfrac{1}{2}]\alpha,\ \sin[n-\tfrac{1}{2}]\alpha,\ -1-\sin \alpha/4)$

It should be noted that $\sin \alpha/4 = (\tau - 1)/2$, tying this set, just like the first, to the golden ratio. Also, if one applies the multiple-angle formulas (and the Pythagorean identity $\sin^2 \theta + \cos^2 \theta = 1$) of Appendix A to this value, one obtains $\cos \alpha/2 = \tau/2$, so the factor $2 \cos \alpha/2$, which appears as the ratio of the x- and y- coordinates of B_n to A_n and of C_n to D_n, is simply τ.

Another set might be generated from the formulas for the coordinates of a pentagonized globoid in Chapter 24, setting $n=3$. The necessary calculations to make all the pentagons regular, however, are very involved, and therefore will not be worked out here.

INDEX

A
Accretion.. 49
Acute 5-pyramoid... 87, 89
Acute angle... 100
Acute dihedral angle.. 87, 89, 100
Acute hexagonal 5-pyramoid... 87
Acute triangle... 90
Akisation.. 49
Alternating axis... 10, 52, 128, 129, 142, 143
Angle defect... 18, 20, 38
Antiapex (of a polyhedron)................................... 69, 70, 78, 91, 92, 101, 104, 105, 109
Antibipyramid.. 70, 79-81, 84, 104, 155
Antiprism.......................ii, 27, 29, 30, 33, 34, 38-41, 64, 65, 71, 73, 80, 81, 106, 111, 116, 121, 127-129
Apex (of a polyhedron)..............49, 62, 69-71, 78-81, 83, 87, 88, 91, 92, 94, 101, 104, 105, 109
Apex (of a pyramid).. 46, 49, 79, 84, 96
Apex (of an isosceles polygon).. 14
Apical polyhedron... 46, 69-71, 104-106
Apical truncation.. 109
Apical vertex.. 71
Apically truncated 5-pyramoid... 109, 123
Apicobasal polyhedron... 69-71, 78, 104, 105, 108, 109
Archimedean solid.............25, 34, 35, 37-39, 41, 45-47, 53, 56, 63, 70, 104, 105, 109, 110, 123
Archimedean truncated cube... 37, 49
Archimedean truncated cuboctahedron... 38, 48
Archimedean truncated dodecahedron... 38
Archimedean truncated icosahedron... 38
Archimedean truncated icosidodecahedron.. 38
Archimedean truncated octahedron... 37, 120
Archimedean truncated tetrahedron.. 123
Archimedes.. 34
Augmentation... 49, 57, 63, 68, 71
Augmented antiprism... 72, 81
Augmented converged antiprism... 81
Augmented pentagonal antiprism.. 71
Augmented polyhedron.. 49
Augmented prism... 72
Axial symmetry... 105, 106
Axially symmetric polyhedron... 53, 65, 68, 69, 105
Axis of rotation.. 6

B

Basal truncation..109
Basally truncated prismoid..109, 123
Base (of a 5-pyramoid)..100
Base (of a polyhedron)...26, 52, 63, 64, 69-71, 73, 74, 78-81, 87-91, 100, 101, 105, 108-113, 116, 117, 119-123, 129
Base (of a prism)..26, 28-30, 39, 64, 112, 113
Base (of a prismoid)..26-28, 30
Base (of a pyramid)...21, 63, 71-74, 76, 78, 79, 81, 84, 90, 96, 99, 106
Base (of a quasiprism)..91, 129
Base (of a tectal polyhedron)..124
Base (of an antiprism)..27-31, 39, 64-66, 81, 128, 129
Base (of an isosceles polygon)..14, 87
Base (of an isosceles trapezoid)..91
Base angle...75
Bi/bigeneral polyhedron...128, 129
Biaugmented antiprism..63, 72, 73
Biaugmented hexagonal antiprism...72
Biaugmented pentagonal antiprism..71, 73, 147
Biaugmented prism...72, 73
Bicropped antiprism...65, 102, 129
Bicupola...124
Bifrustum...155
Bigeneral base...108
Bigeneral roof..108
Bipyramid...63, 71-73, 80, 89, 104, 105
Birotunda..124
Boundary-self-crossing polygon..12, 13, 40
C
Cantellation..48
Catalan solid...105
Center of rotation..6
Central angle..24, 27
Circumradius..25
Circumsphere..25
Closure..4, 8, 11
Components of a vector...136
Composite polyhedron..63, 64, 81, 90, 91, 121, 123
Conceptual dualization..56
Converged antiprism...68, 81, 111, 116, 127
Convergence..68, 111
Convergence point...68
Convex polygon...13

Conway procedure 43
Conway, John Horton 4, 14, 43
Critical truncation 47, 112, 114
Cromwell, Peter R. ii
Cube 22-24, 28, 29, 33, 39, 41, 44, 46, 47, 51-53, 70, 80, 81, 104, 112, 121, 152
Cuboctahedron 37, 39, 46-48, 104, 105, 112, 125
Cumulation 49
Cupola .... ii, 110, 119-121, 123, 124, 127
Cupolarotunda 124
Cyclic group 5, 6

D

Deep truncation 47, 55, 112
Deltahedron 79
Deltohedron 79
Deltoid 79
Descartes, René 18, 20
Diapical polyhedron 69-71, 78, 104, 106, 109
Digonal antiprism 27, 142
Digonal cupola 121, 123
Digonal gyrobicupola 123
Dihedral angle 24, 25, 63, 87, 89, 92, 99, 100, 102, 109
Dihedral group 5, 6, 8
Dipyramid 71
Dodecagon 87
Dodecahedron 23, 47, 52, 53, 155
Dot product 76, 138, 153
Dual polyhedron 23, 47, 54, 56-60, 80, 81, 105, 106
Duality 23, 54
Dualization 16, 58, 70, 81

E

Edge (of a polyhedron)18, 21, 24-26, 35, 38, 48, 49, 55, 56, 59, 64, 68, 71, 75, 79, 81, 84, 87-89, 100, 105, 108, 112, 113, 119, 120, 143, 149, 150, 152
Elongated bipyramid 72, 73
Elongated cupola 122
Elongated pentagonal rotunda 122
Elongated pyramid 72
Elongated square gyrobicupola 35
Elongated square orthobicupola 125
Elongation 64
Equiangular polygon 14, 15
Equilateral triangle 28, 29, 31, 34, 38, 39, 64, 66, 71, 74, 76, 81, 104, 112, 113, 119-122, 127, 151
Euclid 19

Euler formula...20, 35, 49, 56
F
Face (of a polyhedron) 18, 20, 21, 24-26, 34, 35, 38-40, 49, 54-57, 59, 63-65, 68, 70-72, 74, 76, 87, 90-92, 96, 99, 100, 104, 105, 108, 113, 121, 143, 152
Face center..105
Fivefold rotation axis..24, 52, 73, 147, 155
Fold... 8
Fourfold alternating axis...23, 142, 143
Fourfold rotation...5, 16, 17
Fourfold rotation axis..24, 41, 44, 51, 70, 145, 152
Fourteenfold alternating axis...10
Fourteenfold rotation..10
Frustification... 46
Frustum..46, 80
Frustum of a pyramid...26, 46, 90, 111
Full augmentation..49, 50
Fusion...50, 63, 64, 71, 91, 92, 100, 102, 110, 115, 123, 124
G
Geometric vector..136
Globoid..ii, 91, 92, 100-102
Golden ratio..156
Great rhombicosidodecahedron..36, 38, 39
Great rhombicuboctahedron..36, 38, 39, 48
Group (mathematical structure)...4
Group (mathematical)...3
Gyrobicupola..110
Gyrobifastigium..123
Gyrobipyramoid...88, 90, 100
Gyroelongated bipyramid..64, 72, 73
Gyroelongated cupola..122
Gyroelongated hexagonal bipyramid..64, 72
Gyroelongated pentagonal bipyramid...71, 147
Gyroelongated pentagonal pyramid..71, 72
Gyroelongated pentagonal rotunda..122
Gyroelongated pyramid...72, 81
Gyroelongated square pyramid..72
Gyroelongation..64
H
Heptagon..21, 27
Heptagonal 5-pyramid...91
Heptagonal antiprism..27
Heptagonal pentagonized globoid..92

Hermann-Mauguin notation..4
Hermann, Carl...4
Hexagon...34, 37, 48, 109, 114
Hexagonal 5-pyramoid...87, 91
Hexagonal antiprism...65, 66
Hexagonal gyrobipyramoid...8
Hexagonal pyramid..106
Hexagonal quasiprism...65
Hexahedron..22

I
Icosahedron...23, 47, 63, 71, 73, 147
Icosidodecahedron...38, 39, 47, 110, 125
Identity transformation...1, 3, 5, 7, 8
Improper rotation...10
Inradius..25
Insphere...25
International Crystallographic Tables..4
International Notation..4
Intersphere...25
Inversion through a point...7, 91
Irregular antiprism..27, 28
Irregular heptagon..27
Isogonal polyhedron...25, 28, 33, 34, 38, 40, 56, 106, 133
Isohedral polyhedron..25, 28, 33, 40, 56, 106, 133
Isosceles pentagon..52, 78, 80, 87, 91, 92
Isosceles polygon...14, 15, 17, 52, 115
Isosceles trapezoid..65, 66, 79, 91-93, 101, 102, 121, 127, 129
Isosceles triangle..............................i, 28, 39, 64, 65, 71, 74, 79, 104, 105, 112, 113, 119-121
Isotoxal polyhedron..25, 28, 33, 40, 56, 106, 133

J
Johnson cupola..120, 121, 127
Johnson pentagonal rotunda...122
Johnson solid..40, 72, 110, 119, 120, 122, 123
Johnson, Norman..........................ii, 35, 39, 40, 64, 71-74, 81, 88, 110, 119-122, 124
Jordan Curve Theorem..13

K
Kepler-Poinsot polyhedra..12
Kite-quadrilateral...14, 79

L
Lateral edge...112, 114, 115
Lateral face...28, 29, 39, 109, 112-114, 119-122, 129
Lateral face of a prism...26, 34, 64

Lateral face of an antiprism	27, 28, 31, 34, 64

M

Mauguin, Charles-Victor	4
Midpoint	150
Midsphere	25
Mirror line	2, 3, 14
Mirror plane	8-10, 41, 84, 96, 115, 116
Mirror reflection	7-11
Modification procedure	43, 44, 49, 51, 54, 56, 58, 63-65, 68, 142
Monocropped antiprism	119, 121, 127
Monocropped converged antiprism	127
Monocropped hexagonal antiprism	65, 66
Monofrustum of a polykite	80
Monofrustum of an antibipyramid	80

N

Nonconvex polygon	13
Nonconvex polyhedron	63

O

Oblique prism	26
Obtuse 5-pyramoid	87, 89
Obtuse angle	90
Obtuse dihedral angle	87, 100
Obtuse hexagonal 5-pyramoid	88
Octagon	15-17, 36, 44, 45, 48, 49
Octahedron	21, 23, 39, 46, 47, 63, 71, 81, 105
Orbifold notation	4
Order (of a cyclic group)	5, 6
Orthobicupola	110
Orthobipyramid	88, 89, 100, 105

P

Parallel cropping	65, 68, 128, 129
Parallelogram	26
Pentagon	ii, 52, 81, 87-90, 92, 94-96, 100, 101, 104, 105, 114, 122
Pentagon	102
Pentagonal 4-pyramoid	60, 62
Pentagonal antibipyramid	155
Pentagonal antiprism	71, 147
Pentagonal bipyramid	72
Pentagonal gyrobirotunda	110, 125
Pentagonal pyramid	72
Pentagonized globoid	ii, 92, 102, 104, 156
Pentakis dodecahedron	49

Peritruncated bipyramid.. 89, 100
Peritruncated hexagonal bipyramid.. 9
Peritruncated polyhedron... 70
Plane.. 5, 6
Plato... 19
Platonic solid........................19-21, 25, 28, 29, 37-39, 41, 43, 45, 46, 53, 56, 63, 70-74, 80, 92, 104, 105
Polar plane... 58
Polar reciprocation... 58, 60
Polar reciprocation radius.. 58
Polygon... 12, 108
Polygon boundary.. 12
Polygon interior... 12
Polyhedron boundary... 18
Polyhedron interior.. 18
Polykite.. 80, 81
Polytimoid... ii, 80, 81
Principal rotation axis ii, 8, 9, 11, 23, 24, 41, 42, 46, 65, 68, 81, 83, 92, 94, 105, 108, 109, 116, 119, 121, 123, 128, 129
Prism..ii, 26, 29, 30, 33, 34, 38-41, 64, 65, 68, 70, 80, 101, 106, 108, 111-113, 115, 116
Prismatoid..108, 109, 111, 116, 117, 119, 121, 126
Prismoid..26-28, 30, 33, 46, 65, 68, 106, 108, 109, 111, 116, 118
Pseudorhombicuboctahedron.. 35
Pugh, Anthony... ii
Pyramid..46, 49, 63, 65, 71-74, 76, 81, 84, 90, 96, 104, 106, 108
Pyramoid...60, 62, 72, 78, 81, 84, 86, 87, 90-92, 94, 96, 99, 100, 104, 109
Q
Quadrilateral.. 48, 78, 93, 120, 121, 126, 129
Quasiprism.. 65, 66, 91, 92, 100, 101, 128, 129
R
Rectangle.. 26, 27, 39, 105, 113, 119-121, 127, 150
Rectangular prism.. 51
Rectification... 46, 78, 79, 112
Rectified cube... 46
Rectified prism... 112, 113, 116, 117
Rectified prismoid.. 116-118
Reflection... 6
Reflection in a line... 2, 3
Regular antiprism.. 27, 28, 31, 72
Regular decagon... 38
Regular dodecahedron... 22, 92, 104
Regular globoid... 101
Regular hexagon... 38

Regular hexagonal pyramid...... 106
Regular hexahedron...... 22
Regular icosahedron...... 104
Regular octagon...... 15, 45
Regular octahedron...... 24, 28, 29, 72, 104, 152
Regular pentagon...... 38, 122
Regular polygon...... 15, 17, 19, 22, 26, 28-30, 34, 39, 40, 53, 58, 64, 66, 71, 73, 76, 79, 81, 90, 101, 105, 122
Regular prism...... 26, 72
Regular tetrahedron...... 23, 104
Regular-polygon-faced polyhedron...... 68
Rhombic dodecahedron...... 22, 39
Rhombic triacontahedron...... 39
Rhombicosidodecahedron...... 38, 39
Rhombicuboctahedron...... 38, 39, 125
Rhombus...... 22, 39, 89, 93, 104, 105, 112, 113
Right 5-pyramoid...... 87, 89
Right angle...... 100
Right dihedral angle...... 87, 100
Right hexagonal 5-pyramoid...... 87
Right triangle...... 81
Roof (of a polyhedron)...... 52, 69-71, 108, 111, 113, 116, 117, 119-122, 129
Roof (of a quasiprism)...... 129
Rossiter, Adrian...... ii
Rotation...... 5-8, 11, 15
Rotation about a point...... 1, 3
Rotation axis...... ii
Rotational symmetry axis...... 53, 63, 64, 69
Rotocenter...... 6
Rotunda...... ii, 110, 119, 122, 124

S

Scalar product...... 76, 138
Scalene triangle...... 28
Schönflies notation...... 4
Schönflies, Arthur Moritz...... 4, 7
Self-dual polyhedron...... 23
Shallow truncation...... 47, 112, 113
Side (of a polygon)...... 12, 14, 16, 17, 27, 30, 40, 70, 81, 108, 112, 113, 119, 121, 122, 129
Simple polyhedron...... 63, 64, 89, 122
Sixfold alternating axis...... 24
Sixfold rotation axis...... 8, 65
Slant height...... 75
Small rhombicosidodecahedron...... 36

Small rhombicuboctahedron... 36, 39, 48
Snub dodecahedron... 38
Sphere... 25
Square... 15, 19, 26, 29, 34, 35, 38, 39, 44, 56, 64, 66, 80, 81, 104, 105, 112, 113, 119-121, 127
Square bipyramid... 71, 72
Square cupola... 49, 120
Square prism... 26, 28, 121
Square pyramid... 72
Square rotunda... 122
Stellated polyhedron... 12
Stretched dodecahedron... 52
Stretched icosahedron... 73
Stretched octahedron... 73
Strombic pentagon... 52
Strombic polygon... 14, 15, 52, 78, 79
Strombus... 14
Subtectal face... 108, 112
Symbol (of a symmetry group)... 4-8, 10, 11, 23, 24
Symbol (of a transformation)... 5, 8-10
Symmetry... 1, 51, 56, 58
Symmetry element... 6, 14
Symmetry group... i, 3-6, 10, 11
Symmetry of an object... 1, 3
Symmetry transformation... 2-8, 10, 11, 34, 111
Symmetry transformations of a square... 2, 3, 6

T

Tectal polyhedron... 69-71, 105, 106, 108, 129
Tenfold alternating axis... 24, 52
Tetrahedron... 21, 23, 36, 37, 47, 54, 56, 63, 71, 74, 142, 143
Tetratruncated cube... 46
Threefold rotation axis... 23, 24, 41, 44, 51, 52, 70, 81, 109, 145, 147, 152, 153, 155
Total angle defect... 18, 38
Transformation... 1
Transitivity class... 7, 17, 25, 28, 34, 38, 108, 113
Trapezohedron... 79
Trapezoid... 27, 79, 102, 114
Triangle... 19, 27, 34, 35, 39, 56, 66, 71, 72, 78, 81, 87-92, 94, 96, 99-101, 105, 109, 114, 122, 126
Triangular antibipyramid... 81, 121, 153
Triangular antiprism... 28, 81
Triangular bipyramid... 72
Triangular gyrobicupola... 125
Triangular pentagonized globoid... 92

Triangular peritruncated bipyramid..93
Triangular prism..121, 123
Triangular pyramid..72
Trigonometric formulas...134
Truncated..89
Truncated bipyramid...112
Truncated cube...36, 45, 112
Truncated cuboctahedron..48
Truncated icosidodecahedron..38
Truncated octahedron...37, 47, 112, 120
Truncated prism...112
Truncated pyramid...109, 123
Truncated tetrahedron..37, 109
Truncation...37, 39, 44-47, 49, 55-57, 65, 68, 70, 78-80, 109, 112-114
Truncation plane..79
Twelvefold alternating axis...65
Two-plane parallel rule..126, 139-141
Twofold rotation axis......................................8-10, 23, 41, 42, 44, 51-53, 104, 105, 145, 147, 150, 152, 155

U

Uni/bigeneral polyhedron..108-110, 119, 126
Uni/bigeneral tectal polyhedra..115
Uni/bigeneral tectal polyhedron...121-123
Uni/unigeneral polyhedron...108-111, 116, 123
Uni/unigeneral tectal polyhedra...112, 115
Uniaxial stretching..43, 51-53, 70, 71, 73, 104
Uniaxially stretched Archimedean solid...53
Uniaxially stretched Platonic solid...53
Uniform polyhedron..40
Unigeneral base...108
Unigeneral roof..108

V

Vector...83, 136, 153
Vector algebra..136
Vector geometry...136
Vertex (of a polygon).................12, 14, 15, 17, 18, 27, 29, 30, 40, 65, 71, 81, 87, 96, 102, 105, 113, 119
Vertex (of a polyhedron).......18, 21, 25, 34, 35, 38, 46, 48, 49, 54-57, 59, 64, 65, 68-70, 74, 75, 79, 91, 94,
   105, 108, 112, 115, 122, 143, 147, 150, 152, 153
Vertex (of a polyhedron) 34
Vertex (of a pyramid)..84, 96
Vertex (of an antiprism)...31
Vertex (of the base of a polyhedron)..27
Vertex angle...38

Vertex figure.. 35, 38
Vertex rotation plane... 64, 68-70, 78, 79, 81, 83, 99, 102, 117, 122, 142, 153
Vertex rotation plane based labeling... 69, 74, 83, 94, 147, 153

W

Wikipedia... 34, 64, 119, 120

Z

Zalgaller, Victor 40

www.ingramcontent.com/pod-product-compliance
Lightning Source LLC
Chambersburg PA
CBHW081237180526
45171CB00005B/448